# Cybersecurity for Entrepreneurs

# Cybersecurity for Entrepreneurs

BY

**GLORIA D'ANNA AND ZACHARY A. COLLIER**

Warrendale, Pennsylvania, USA

400 Commonwealth Drive
Warrendale, PA 15096-0001 USA
E-mail:    CustomerService@sae.org
Phone:    877-606-7323 (inside USA and
          Canada)
          724-776-4970 (outside USA)
FAX:      724-776-0790

**Library of Congress Catalog Number 2023938108**
**http://dx.doi.org/10.4271/9781468605730**

**ISBN-Print  978-1-4686-0572-3**
**ISBN-PDF    978-1-4686-0573-0**
**ISBN-ePub   978-1-4686-0574-7**

**To purchase bulk quantities, please contact: SAE Customer Service**

E-mail:    CustomerService@sae.org
Phone:     877-606-7323 (inside USA and Canada)
           724-776-4970 (outside USA)
Fax:       724-776-0790

**Visit the SAE International Bookstore at books.sae.org**

**Publisher**
Sherry Dickinson Nigam

**Development Editor**
Amanda Zeidan

**Director of Content Management**
Kelli Zilko

**Production and Manufacturing Associate**
Brandon Joy

## Dedication

*This book is dedicated to our friend Peter the Salesman. Thank you for all of your stories. And, after reading this book and implementing the lessons learned, may you have only successful cyber days as an entrepreneur!*

# Table of Contents

Acknowledgments    xv

**CHAPTER 1**

Cybersecurity: The Sunscreen of the
Information Technology World    1

| | | |
|---|---|---|
| 1.1. | Cybersecurity: Don't Get Burned! | 1 |
| 1.2. | A Gap in Cybersecurity Education | 3 |
| 1.3. | So Why Should You Care? | 4 |
| 1.4. | Who Is Peter the Salesman? | 4 |
| 1.5. | What Will You Learn in this Book? | 6 |
| References | | 7 |

**CHAPTER 2**

Cybersecurity Advice from the
Angel and the Devil    9

| | | |
|---|---|---|
| 2.1. | Peter the Salesman Meets the Angel and the Devil | 9 |

**CHAPTER 3**

Securing Your Communications: E-mail,
Web, and Phone    15

| | | |
|---|---|---|
| 3.1. | Introduction | 15 |
| 3.2. | "Left of Bang" Cybersecurity Awareness | 16 |

3.3.  Communication Dos and Don'ts                          18

3.4.  E-mail                                                 19
      3.4.1.  Limitations of E-mail Security                 19
      3.4.2.  Secure E-mail                                  20
      3.4.3.  Phishing and Spear Phishing                    21

3.5.  Web Safety                                             21
      3.5.1.  Cloud Security                                 21
      3.5.2.  Web Security                                   22
      3.5.3.  Web Tracking                                   22
      3.5.4.  Bad Sites and Links                            23

3.6.  Phone                                                  23
      3.6.1.  Recommended Phones                             23
      3.6.2.  Secure Phones "As Seen on TV"                  24
      3.6.3.  Voice, Text, and Messaging                     24
      3.6.4.  Cars, Events, and Overseas Travel              25
      3.6.5.  PACE Communications                            25

3.7.  Post Quantum Cryptography                              26

3.8.  Conclusions                                           26

References                                                   27

**CHAPTER 4**

# Protect Your Financial Transactions Now! Cybersecurity and Finance for the Entrepreneur    31

4.1.  Introduction                                          31
4.2.  A Little Background on Data Breaches that
      an Entrepreneur Should Consider                        32
4.3.  How to Keep Your Finances Safe                        34
      4.3.1.  Identity and Access Management (IAM)           34
      4.3.2.  Data Encryption                                35
      4.3.3.  Business Continuity (Backup and Restore!)      35

4.4.   The Cloud, Data, and Software-as-a-Service    36

4.4.1. API Security    38

4.5.   Credit Card Processing Compliance and Standards    39

4.6.   Conclusions    41

References    41

CHAPTER 5

# Who Needs a VPN?    43

5.1.   Introduction    43

5.2.   What Is a VPN?    43

5.3.   But No One Is Spying on Me    47

5.4.   Do I Need to Use a VPN When Surfing the Internet?    51

5.4.1. Wow! It Is Hopeless!    52

5.4.2. Threats to VPN Traffic Are Everywhere!    52

5.5.   What Features Matter Most in a Modern VPN Service?    53

5.6.   It Is 2023, What VPN Fits the Bill?    54

5.7.   Conclusions    55

References    57

CHAPTER 6

# Securing Your IoT Devices    59

6.1.   Introduction    59

6.2.   Reduce Your Attack Surface    60

6.3.   Keep Your Devices Updated    61

6.4.   Cutting Out the "Middle Man"    63

6.5.   Practice Good IoT Cyber Hygiene    66

6.6.    Conclusions                                          69

Reference                                                    70

**CHAPTER 7**

Product Security for Entrepreneurs
Selling Digital Products or Services          71

7.1.    Introduction                                         71

7.2.    Flaws in Digital Products Can Be Expensive           72

7.3.    Shifting Security Earlier                            73

7.4.    A Basic Security Approach                            75

7.5.    Threat Modeling                                      77

7.6.    Testing                                              81

7.7.    Sustainment                                          85

7.8.    Conclusions                                          89

References                                                   90

**CHAPTER 8**

Strategic Startup in the Modern Age:
Cybersecurity for Entrepreneurial
Leaders                                        93

8.1.    Introduction                                         93

8.2.    Modern Entrepreneurial Strategies                    95

          8.2.1. Effectuation                                96

          8.2.2. Design Thinking                             96

          8.2.3. Systems Thinking                            97

          8.2.4. Entrepreneurial Thinking                    98

8.3.    Modern Entrepreneurial Tools                         99

          8.3.1. Business Model Canvas                       99

          8.3.2. Lean Startup                                100

          8.3.3. Customer Development                        100

8.4.  Modern Entrepreneurial Networking            101
    8.4.1. Entrepreneurial Ecosystems               101
    8.4.2. Ecosystem Builders                       102

8.5.  Value of Entrepreneurial Strategies, Tools,
    and Networking in the Digital Age          103

References                                          105

**CHAPTER 9**

# Cyber Law for Entrepreneurs    107

9.1.  Introduction                                  107

9.2.  Federal Laws, Executive Orders, and
    Regulations                                108
    9.2.1. Federal Laws and Regulations             108
    9.2.2. Executive Orders                          113

9.3.  State Laws, Regulations, and Executive Orders 115
    9.3.1. Data Breach Laws                          115
    9.3.2. Minimum Standard and Reasonable Data
        Security Measure Laws                     118

9.4.  European Union and International
    Requirements                                121

9.5.  Practical Considerations: How Cyber Law
    Can Directly Impact Your Business          123
    9.5.1. General Recommendations                  123
    9.5.2. Cybersecurity Terms in Contracts         124

9.6.  Conclusion                                    124

**CHAPTER 10**

# Cyber Economics: How Much to Spend on Cybersecurity    127

10.1.  Introduction                                 127

10.2.  Value of Your Product or Service             128

10.3.  Cybersecurity as a Cost Center versus a
Profit Center                                                130

10.4.  How Much to Spend: Common Economic
Measures of Cybersecurity Spending            132

    10.4.1.  Return on Investment (ROI)              132

    10.4.2.  Risk-Based Return on Investment for
        Cost Center Spending                       134

    10.4.3.  Delayed Net Present Value and
        Catastrophic Cybersecurity Incidents    139

10.5.  Estimating Costs, Benefits, and Other
Information                                                   144

10.6.  Conclusions                                            145

**CHAPTER 11**

# Cyber Insurance for Entrepreneurs    147

11.1.  Introduction                                           147

11.2.  What Is Cyber Insurance?                               151

11.3.  When and How Do I Buy Cyber Insurance?  152

    11.3.1.  When Should I Buy Cyber Insurance?   152

    11.3.2.  How Do I Get Cyber Insurance and Who
        Should I Contact?                          153

    11.3.3.  How Do I Apply for Cyber Insurance?  153

11.4.  What Are Some Controls That Would
Be Important?                                             155

    11.4.1.  Network Security Vulnerabilities      155

    11.4.2.  Email Security                             155

    11.4.3.  Internal Security Controls             156

    11.4.4.  Backup and Recovery                      157

    11.4.5.  Phishing                                    158

11.5.  What Are Some Important Contractual
Aspects to Know about the Insurance Policy? 160

11.6.  What Are Some Important Parts of the
Insurance Policy to Pay Attention To?          160

11.7.   First- and Third-Party Insuring Agreements     162
    11.7.1.  First Party     162
    11.7.2.  Third Party     165
11.8.   Conclusions     167
References     168
    Disclaimer     168

**CHAPTER 12**

## Cyber Resilience for Entrepreneurs     169

12.1.   Protection versus Performance     169
12.2.   Introducing Resilience     170
12.3.   Holistic Approach     173
12.4.   Resilience as a Cycle     175
12.5.   Design Principles     178
12.6.   Taking Action     180
12.7.   Conclusions     182

**CHAPTER 13**

## Cybersecurity for Entrepreneurs... and Beyond     183

13.1.   So What Have We Learned in this Book?     183
13.2.   Epilogue: Peter Looks toward the Future     186

About the Authors     189
About the Illustrator     205
Index     207

# Acknowledgments

Zach and Gloria would love to thank SAE International who made this book possible. Thank you to SAE's Sherry Nigram who believed in this project from the onset! Thank you to Amanda Zeidan who has been with us every step of the way.

We would like to thank our authors, who also went on this adventure with us: Simon Hartley, Chris Sundberg, Dennis Vadura, Peter Laitin, Kenneth Crowther, Dale Richards, Samantha Steidle, Jennifer Dukarski, Ariel Pinto, Luna Magpili, Howard Miller, and Paul Roege.

A big thank you, also to our SAE G-32 Cyber Physical Systems Security Committee for the cyber expertise—as we move from large companies to entrepreneurs for this endeavor!

And, we would love to thank Phillip Wandyez, our illustrator, who helped bring Peter the Salesman to life!

And finally, we would like to thank Gloria's two friends Peter, who when added together, and then embellished, served as the inspiration to help create Peter the Salesman.

As you, the reader, work through your cybersecurity issues, ask yourself, "What Would Peter Do?"

May you enjoy our book.

"Have a Safe Cyber Day!"

XXOOXX

Gloria and Zach

# 1

# Cybersecurity: The Sunscreen of the Information Technology World

Gloria D'Anna and Zachary A. Collier

Illustrated by Phillip Wandyez.

## 1.1. Cybersecurity: Don't Get Burned!

In your early years, you probably ran outside without any sunscreen. And you probably got a tan. Or maybe you got a sunburn and got freckles on your shoulders. Or maybe you really got burned and your mom got you some Aloe Vera to put on your burn, and for a few days, it was painful and itchy.

Then, as a result, maybe you got more cautious, and the next time you went outside to play, you put on some sunscreen (at your mother's insistence).

Or maybe you ignored your mother, continued to get burned, and ended up with skin cancer in your later years. And, after having skin cancer removed, with a hefty medical bill, decided to adhere to your doctor's advice and wear sunscreen. And maybe you even wear special sun protection factor (SPF) clothing. Plus, a big floppy hat.

And then you go outside when it is hot out, with a hat, sunscreen, and SPF clothing from head to toe, and you get overheated when you go outside. Or maybe you just decide to skip the sun and sit on the couch during the daytime and only go outside once the sun has gone down.

In all these scenarios, there is a known risk that you take with the sun and also with the sunscreen.

So it goes with cybersecurity.

Let's say that as an entrepreneur you are so ill informed about cybersecurity that you just end up doing nothing. This is like going outside as a child without sunscreen, and you get burned. Your email gets hacked. Your customer lists get hacked. But you did not spend any money.

On the other end of the spectrum, maybe you are so paranoid about the sun that you sit on the couch all day and only go out at night. Equivalent to not having an email account and not having a website and not having anything except a phone number. And you have the SPAM function on your phone turned on. This is great in terms of protecting yourself from potential threats, but it does not really work for your sales. Cybercriminals cannot reach you, but neither can your customers!

Maybe there is a sweet spot somewhere in between, where you can be well informed and protected against potential threats, while, at the same time, leverage the latest and greatest technologies to help your company grow and thrive.

You want to be informed about cybersecurity for your business. And that is what this book is all about!

We have put this book together for you. And after reading it, you should be better equipped to determine how much risk you want to take, and how much money you want to spend during the early stages of your company when cash flow is tight and you might look for things that are "free" yet secure enough. Then as your business grows, and your cyber risk profile grows, you can add on to your cybersecurity. But you should think about it and make good, risk-informed decisions.

Our goal here is to get you to think. In particular, we want you to think about how cybersecurity fits within the overall picture of your business, along with all of the other important things you do as an entrepreneur or small business owner.

## 1.2. A Gap in Cybersecurity Education

Many universities offer some type of entrepreneurship education, whether it be classes on the topic or even degree programs related to starting and running your own business. Traditional entrepreneurship classes cover topics like how to understand customer pain points, how to identify good business opportunities, how to write a business plan, how to acquire funding for your business, and how to take an idea and turn it into a successful new venture. Of course, these are all essential parts of entrepreneurship. But is that all there is to entrepreneurship, or could it be supplemented somehow?

In an entrepreneurship class, a student often learns a little bit about many different business disciplines. As an entrepreneur, you are your own chief marketing officer, your own chief financial officer, and your own chief operating officer, plus the chief of accounting, research and development, and human resources. And so you learn a little bit about marketing, finance, operations, accounting, product design, and management.

But as it turns out, you are also your own chief information security officer. And for some reason, entrepreneurship classes typically gloss over that part.

We stumbled across this realization somewhat accidentally. Zach was teaching an entrepreneurship class at the time at Radford University and had invited Gloria to speak with his class as part of a series of guest speakers. Specifically, Gloria talked about cyber-physical systems security and the SAE G-32 Committee (of which

Gloria co-chairs and Zach is a member). During the guest talk, Gloria spoke about the importance of keeping one's computing and information technology infrastructure secure, especially as an entrepreneur who is just starting out and does not have much money to spend on security. How do you know that the tools and platforms that you are using to run your business are secure?

After that class, Gloria said, "Zach, we really should write a book about this." We decided that there was an important gap in entrepreneurship education related to cybersecurity so we set out to write this book.

## 1.3. So Why Should You Care?

At this point, you might be thinking, "Okay, so what? Don't hackers just target big companies with all the money and data? Nobody cares about my little mom-and-pop shop."

Not true! According to Verizon's 2022 Data Breach Investigations Report [1], 43% of data breaches involve small and medium-sized businesses, and 61% of all such businesses have reported at least one cyberattack during the last year. One in 323 emails sent to businesses is malicious [2].

And the impacts of a successful attack can be catastrophic for your entrepreneurial venture. The average cost of a cyberattack is $120,000, yet 54% of small and medium-sized companies do not even have a plan about what to do should a cyberattack occur [3].

A cyberattack can easily put your new venture out of business. You do not want that to happen (and neither do we!), so we wrote this book to help you build a baseline of cybersecurity knowledge that you can use in your business.

## 1.4. Who Is Peter the Salesman?

Cartoonist Ashleigh Brilliant once observed that "It could be that the purpose of your life is only to serve as a warning to others." Peter the Salesman is a character that was created as our warning to you about what not to do when it comes to cybersecurity (Figure 1.1).

**FIGURE 1.1**  Meet Peter the Salesman.

Illustrated by Phillip Wandyez.

Back in 2020, Gloria and Zach wrote a paper titled, *A Holistic Approach to Cyber Physical Systems Security and Resilience* [4]. In that paper, Gloria introduced a character called Peter the Salesman. Peter was a fictional character based on a few of her friends. Peter always seemed to have a bad cyber day. Everything he touched just had a way of going awry. For example, Peter's credit card information gets leaked. Then the self-driving taxi he is taking is compromised to travel at unsafe speeds. Later on, the navigation systems in the plane he is taking during a flight malfunction and take him to the wrong city.

At one point, there was some debate about whether Peter's bad cyber day would lead to his untimely demise. Gloria wanted to kill him off, but Zach was very upset about it. "You can't kill Peter off due to a cyber accident!" he decried. So Peter is alive and shows up throughout the book chapters. We show through the misadventures of Peter the Salesman, as he is starting a new company, what not to do as he tries (and inevitably fails) to navigate the cybersecurity pitfalls of doing business in the digital age.

The best way to learn from mistakes is to let other people make them.

## 1.5. What Will You Learn in this Book?

Our goal with this book is to make it easy (and maybe a little bit fun) to read for entrepreneurs. It is a starter book. You do not have to possess a PhD in Computer Science to understand the concepts we cover in the upcoming chapters.

We have reached out to our cybersecurity expert colleagues from industry and academia to write focused, topical chapters that provide practical guidance on cybersecurity topics that are relevant from the perspective of entrepreneurs and small business owners. The authors know their subjects so well that they can make things easy to understand. We believe that we have succeeded in that.

The book is divided into roughly two sections. The first half covers cybersecurity basics like how to secure your communications, financial transactions, data, and devices. There is a chapter on secure product development as well, in case your entrepreneurial business involves products that must themselves be secure. If you are opening a cupcake bakery, you could probably skip that chapter, unless your cupcakes connect to the Internet somehow. The second half of the book is more about the business aspects of cybersecurity. Topics are covered such as entrepreneurial planning for cybersecurity, law, economics, insurance, and cyber resilience.

We hope that by reading this book, you feel better informed about your role as chief information security officer of your new company (Figure 1.2).

**FIGURE 1.2**    Don't get burned.

Illustrated by Phillip Wandyez.

# References

1. Verizon, "2022 Data Breach Investigations Report," 2022, accessed August 27, 2022, https://www.verizon.com/business/en-gb/resources/reports/dbir/.

2. Shepherd, M., "30 Surprising Small Business Cyber Security Statistics (2021)," Fundera (by NerdWallet), 2020, accessed August 27, 2022, https://www.fundera.com/resources/small-business-cyber-security-statistics.

3. InsuranceBee, "Cybercrime Survey Reveals SMB Owners are Unaware and Unprepared," 2018, accessed August 27, 2022, https://www.insurancebee.com/blog/smb-owners-unprepared-for-cybercrime.

4. DiMase, D., Collier, Z.A., Chandy, J., Cohen, B.S. et al., "A Holistic Approach to Cyber Physical Systems Security and Resilience." in *Proceedings of IEEE/NDIA/INCOSE Systems Security Symposium*, Crystal City, VA, 2020.

# 2

# Cybersecurity Advice from the Angel and the Devil

Gloria D'Anna and Zachary A. Collier

## 2.1. Peter the Salesman Meets the Angel and the Devil

Peter the Salesman was feeling overwhelmed. He was starting his business and was busy with all of the things that he had to do to get his company up and running. Designing prototypes for his new product, getting feedback from customers, talking to vendors… There just were not enough hours in the day. And of course, other than the money he had saved up, he did not have any other sources of revenue coming in yet.

The last thing on his mind was cybersecurity. Peter's head was full of business ideas but nothing on cybersecurity.

Of course, he had read stories in the news about companies getting hacked, but he did not think that could happen to him.

All of a sudden, Peter began to hear voices in his head. He whacked his hand on the side of the head, but they were still there. "I must be overworking myself," Peter said to himself.

There was this lovely voice on his left shoulder. An angelic voice. "Peter, you really should think about cybersecurity…" Then on his right shoulder, there was another voice, but this one was a bit squeaky and raspy, perhaps from smoking too many cigarettes. It was saying, "Peter, that cybersecurity is junk bunk… just ignore it."

Peter decided to go for a walk to get some fresh air and clear his head. When he got back to the office, he was feeling better and the voices had disappeared.

He needed to set up some business pages for his new company on social media, so he got back to work. "Hmmmm…," Peter thought to himself, "this website is asking me to choose a password. I am always forgetting things, so I should try to come up with something that I can remember." After some thought, a lightbulb went off in Peter's mind, and he entered: "password." Peter felt very clever (Figure 2.1).

**FIGURE 2.1**    Peter has a bright idea.

Illustrated by Phillip Wandyez.

"You better not use that password Peter," the angelic voice said.

"Oh no, the voices are back," Peter thought to himself.

"*Password* is one of the most commonly used passwords across the world," continued the angel. "Don't you think the bad guys know this? They could guess your password and gain access to your account. You are better off using a combination of letters that are both upper and lower case, numbers, and special characters."

Peter decided to create a strong password following the angel's advice.

"You'll never be able to remember that password with all those special characters! You better write it down on a sticky note and put it on your computer monitor in case you forget," said the devil.

"Hmm, that sounds reasonable," thought Peter.

"Peter, don't do it! Leaving your password written down and out in the open for everyone to see is going to lead to someone stealing it," the angel said. "In fact, don't write it down at all."

Peter listened to the angel and did not write down his password on a sticky note.

The devil was clearly miffed. He dug his heel into Peter's shoulder and caused quite a pain in Peter's neck.

Once Peter had logged in, the website asked him if he wanted to enable two-factor authentication. "I do not even know what that is," said Peter to himself. "Maybe one of these voices can help me."

"Two factors? Sounds like a lot of extra steps to me," the devil said. "You don't want to let all of this extra cybersecurity stuff slow you down... time is money!"

"You should enable it," said the angel. "It will verify that you are who you say you are, adding an extra layer of security when you sign in. Think of security like a delicious, multilayered cake. The more layers you have, the better."

"You know what else has a lot of layers? Onions! And onions make you cry," interrupted the devil.

Chopping up onions did always bring a tear to Peter's eyes.

"Peter, what if one of your defenses doesn't work? The idea is to have multiple layers of defense, just in case one fails. Think about driving your car—you wear your seatbelt, and you also have airbags, you follow the rules of the road, and so on. These are all layers of protection that you use when you drive a car, so why shouldn't you have layers of protection for cybersecurity?"

"Blah, blah, blah," said the devil.

"Geez, I need a vacation," thought Peter.

On and on it went like this, with the angel on one shoulder, and the devil on the other, arguing with each other. Peter kept a list of the advice that the angel and the devil gave about various cybersecurity topics (Table 2.1).

**TABLE 2.1** Cybersecurity advice from the angel and the devil.

| Angelic (GOOD) advice | Devilish (BAD) advice |
| --- | --- |
| Keep your software, including operating systems, up to date. | Do not worry about those pesky security updates! |
| Be careful about connecting to public Wi-Fi. | Just connect to whatever Wi-Fi, what is the worst that could happen? |
| Be careful about using unknown USB flash drives. | Just use any random USB drive you find—those files are not going to transfer themselves! |
| Use an antivirus software and make sure to keep it up to date. | Nah, do not worry about it. |
| Use two-factor authentication to secure important accounts. | Skip it, too many extra steps. |
| Follow best practices when it comes to making strong passwords. | Just use the same simple password across all of your accounts. |
| Be careful about suspicious emails, do not click on links that appear suspicious, and be careful about downloading attachments from untrusted sources. | Click on those suspicious email links... who knows, maybe you really did win a free giveaway. |
| Be careful about suspicious websites, make sure the websites you visit use the "HTTPS" prefix. | Surf the net with reckless abandon. |
| Lock your devices when you are not using them, including laptops, phones, and tablets. Do not leave them unattended. | Leave your devices laying around in public. Logging out and then logging back in is so much work! |
| Keep your data backed up, and back it up regularly. | *Yawn...* BORING! |
| Use encryption when storing and transmitting sensitive data. | Encryption? Sounds like math. Skip it. |

Eventually, the voices were in a heated argument about something called "Product Security."

The angel said, "Peter, you need to be transparent. You need to let your customers know that you are on a product security journey. Communicate your security context as clearly as possible."

Peter, thought to himself, "Such a pleasing angelic voice. But what the heck is a product security journey?"

The devil in his raspy voice argued back, "Hide and obscure security. Avoid the topic of product security. Exaggerate existing security."

The angelic voice kept on, "Explain to your customers that security is a shared responsibility; provide secure deployment guidance that helps complement existing security measures of the product."

The devil argued vehemently, "Tell the customers that they are solely responsible for the security of their operations, even if they use your product. Pretend that your digital product is completely secure and tell the customers that they don't need to be concerned about security."

Peter thought to himself, "Oh, I understand the devil here. It is the customers' problem. I really like that solution."

The angel shined a very bright LED flashlight into Peter's eyes when he thought that. Peter thought he was going blind. "Oh, maybe that was the wrong thought," Peter said. "I should probably listen to the angel."

Peter wanted to do the right thing when it came to cybersecurity, but he just had no idea where to start.

"THIS IS SO DIFFICULT!!!" Peter yelled in frustration.

And just like that, a halo appeared. Then the angel appeared before Peter, holding a book titled *Cybersecurity for Entrepreneurs*.

The devil tried to set the book on fire, but Peter blew out the flame.

Peter opened the book and started reading...

# 3

# Securing Your Communications: E-mail, Web, and Phone

Simon Hartley

## 3.1. Introduction

Communication with stakeholders is the heart of a business. We all need to stay in touch with the folks who are helping us grow, whether customers, prospects, partners, investors, or reviewers. An initial contact becomes a meeting, a demo, and a sale, or plants a seed for the future in a self-reinforcing spiral of success.

Business communications range from traditional events, face-to-face meetings, and calls to a myriad of electronic channels, from webinars and social media to e-mails, instant messaging, and texts. Post-COVID, there has been a shift toward more electronic contacts. The ideal there is to foster communications and avoid extremes like ignoring unfamiliar channels or forcing everything to be a phone call. Worst of all might be inadvertently sharing too much with scammers pretending to be someone else.

Maintaining confidentiality, integrity (no tampering), and availability of digital communications (the so-called "CIA" Triad) need not be a "007" level of effort (although that is the international dialing code for Russia). It can be achieved affordably and easily with a thoughtful set of security principles, cloud-based tool choices, and the setups detailed below. Ease of use is a key theme since it is the security that you use that matters most, not that "shelf ware" service that you are paying for but never set up.

## 3.2. "Left of Bang" Cybersecurity Awareness

Cybersecurity awareness and basic good practices can go a long way to keeping you "left of bang" [1]. This is the term the United States (US) Marine Corps uses for a mindset that avoids walking into ambushes or roadside bombs. Let us quickly review some of the proactive points to avoid the cybersecurity "bang" of a data breach, ransomware, or intellectual property (IP) theft. Right of bang is the stress, time, and cost of engaging legal and other remediation experts.

Your greatest tool is your awareness. If you hesitate to put your credit card into a homemade-looking card reader at a sketchy gas station, you might also question why Amazon, UPS, or the IRS are asking you for sensitive information or to click on "their" link. No "anti-phishing" tool looking to warn you of scammers is 100% accurate. That request to urgently transfer $10,000 overseas might actually be from your supplier's account (controlled by bad guys) but with changed banking information. It is worth confirming similar requests via another channel if the stakes are high for a particular request.

Who makes and maintains the tools you are using? Are you working with reputable companies, staff, and investors behind them? Can you verify bona fides, or are their employees anonymous? Do they use marketing hyperbole like "un-hackable," "un-traceable," and "government-grade security"? They may operate from countries with authoritarian regimes, have no gaps between government and commercial interests, and have ties to hackers for hire or even criminal gangs. An untrustworthy security tool is worse than no tool at all.

Some software security tools may be just a "side hustle" from a solo developer or a loose group of part-time developers, where the risk is that as they get busy or travel in their day jobs then you lose the ability to reach them for updates and support. It is worth considering how your own company itself might be perceived by customers, partners, or investors carrying out "Know Your Client" (KYC) types of due diligence by looking at the information (or absence of information) currently on your website, LinkedIn, or similar public profiles.

**FIGURE 3.1**   Peter visits an insecure website.

Illustrated by Phillip Wandyez.

*After learning about the importance of strong passwords, Peter figured that there was probably a lot of other cybersecurity information that he did not know. "I should probably brush up on this cyber stuff," he thought to himself as he searched for cybersecurity tips on the Internet. Clicking on a search result, his browser alerted him that the website he was attempting to visit was not secure. "Not secure? How is that even possible? This website is literally about cybersecurity!" he said as he ignored the prompt and continued to the site, which promptly downloaded malware onto his computer.*

## 3.3. Communication Dos and Don'ts

Communications security for e-mail, websites, and phones do not rely on a single tool or choice, but rather a thoughtful overall selection and setup. Security is not defined by the strongest of the tools that we have but by the weakest among them, which we already know may be ourselves or our co-workers! The list below is a summary for an entrepreneur or small business owner to consider, even as choices evolve over time.

Stakes around a cybersecurity "bang" can be much higher in certain organizations like government and C-suites and for high-net-worth families (HNW) as the UK Prime Minister recently discovered [2]. Their tools and configuration follow similar principles but are maintained by in-house teams or are outsourced to specialist boutiques like **Glacier Security** [3].

1. Use new **Apple, Pixel**, or **Samsung** smartphones and data plans purchased from a reputable source rather than at a kiosk in an airport, railway station, or port.

2. Use a password manager like **Dashlane** [4] or **1Password** [5] to generate complex passwords to make their use easy.

3. Use two-factor authentication applications (apps) (2FA) for cloud access. Apps are preferred to texts since the latter are vulnerable to subscriber identity module (SIM)-swapping attacks [6].

4. Use encryption options for all stored or transmitted data, where possible, but especially for sensitive data such as designs, finance, or legal documents.

5. Use encrypted messaging apps like **Signal** [7] or **Wickr** [8] behind a trusted virtual private network (**VPN**) (Chapter 5) rather than e-mail, text messages, or voice calls for sensitive data.

6. Use websites with Hypertext Transfer Protocol Secure (https) encryption enabled, i.e., the secure "lock" symbol is visible.

7. Use automated updating (patching) of tools, where available.

8. Use automated backup and encryption for all data.

9. Use continuous monitoring tools, e.g., **Lookout** [9] or **Zimperium** [10] on a smartphone.

10. Use opt-outs of preview displays, location, and data tracking, where available.

11. Use hardware-encrypted and personal identification number (PIN)-enabled external storage devices like **Apricorn** [11] when traveling with sensitive data.

12. Set up multiple different ways of staying in touch and carry batteries in case of outages and emergencies.

13. Avoid using third-party Wi-Fi directly. See the chapter on **VPN** (Chapter 5).

14. Avoid using third-party USB thumb drives, chargers, and cables that may infect your equipment and/or steal your data.

15. Avoid leaving your equipment unattended, especially at industry conferences and/or when traveling overseas. You may be a target for industrial or nation-state espionage that requires physical access to your device [12].

For more in-depth information, the **Center for Internet Security** [13] (CIS) has one of the easiest-to-use checklists for Information Technology, including secure communications. The Department of Homeland Security **Cybersecurity and Infrastructure Agency** [14] (CISA) also offers good advice for small businesses.

## 3.4. E-mail

### 3.4.1. Limitations of E-mail Security

E-mail is a key communication platform for everyday business. It also ties to our contacts, calendars, locations, attachments, and links. While data within the body of e-mails can be encrypted, the metadata around whom we are corresponding with, when, and by what route is typically open. Secure messaging platforms (behind a trusted **VPN**) offer greater security than e-mail. Their security is also greater than text messages and phone calls. The key to this is end-to-end encryption (E2E), which regular e-mail, texts, and phone calls do not have.

Today's most common e-mail platforms are **Microsoft Outlook** and **Google Gmail**. Both are full featured. However, security is not their default option. All the major technology companies surveil their customers, monetizing their locations, data, calendars, and contacts for internal benefit, and/or sell data to third-party data brokers and advertising networks. ADint [15] or OSint [16] (advertising or open-source intelligence gathering) from these sources allows stalkers, competitors, activists, and criminal gangs to target us.

As far back as 2012, US Supreme Court Justice Sonia Sotomayor described the "chilling effect" of such digital surveillance [17] "monitoring generates a precise, comprehensive record of a person's public movements that reflects a wealth of detail about her familial, political, professional, religious, and sexual associations.... Disclosed in data will be trips the indisputably private nature of which takes little imagination to conjure - trips to the psychiatrist, the plastic surgeon, the abortion clinic, the AIDS treatment center, the strip club, the criminal defense attorney, the by-the-hour motel, the union meeting, the mosque, synagogue or church, the gay bar and on and on."

Google recently announced that they will delete past appointments to medical care facilities from our accounts for privacy reasons [18], which begs the question of why the company was monitoring such appointments in the first place!

### 3.4.2. Secure E-mail

Security features like **Pretty Good Privacy** [19] (Open PGP) can be layered into Outlook or Gmail e-mails. However, they are notoriously difficult to use [20]. They can potentially impact relationships with stakeholders, where the conversation focus can shift from your business to how to use your unfamiliar security tools. Switzerland's **Proton Mail** [21] offers similar e-mail features to Microsoft and Google. However, PGP security is enabled by default and much easier to use compared to bolt-on tools. Proton Mail, Microsoft, and Google can all be tied to your company domain name, so you can focus on building your business brand rather than advertising the brand of your infrastructure provider!

### 3.4.3. Phishing and Spear Phishing

The largest security issue with e-mail is phishing, where the name of a trusted organization or colleague is used to trick users into clicking on a malware link, or into disclosing sensitive information. E-mail addresses, like website names, phone numbers, and/or text messages can all be faked or spoofed. If you are in doubt, do not act on the communication.

Many phishing e-mails supposedly from the IRS, Amazon, or UPS are easy to spot with spelling mistakes, bad grammar, or unusual requests such as buying gift cards. Why reply to messages from obviously suspect accounts like "fedex666" or "AmericanExpress88"? Care must be taken when they are well written, supposedly from a trusted family member, friend, or colleague, and the ask is around IP, financial, or legal documents.

Also, avoid providing the answers for your top five security questions on popular social media quizzes like: "What was the name of your favorite dog? Where did you meet your spouse?" that may be designed simply to harvest the name of your hometown, pet, mother's maiden name, and so on, along with innocuous filler questions.

## 3.5. Web Safety

### 3.5.1. Cloud Security

The cloud is just your data on someone else's computer! To make it easy for you to access (but not everyone else), use a password manager to keep track of complex passwords like "YhAByGTXjzg3B17VWrQs" and use two-factor authentication, which means (for example) entering a password in the browser and then confirming access via your smartphone. Just as for local devices, verify that your data in the cloud are encrypted and regularly backed up and that vendor infrastructure is kept up to date.

### 3.5.2. Web Security

The key to security on the Web is something called "https," where the "s" stands for secure communications. Older and/or malicious websites use "http" (with no "s") and lack the small lock symbol next to the Web address in the browser. When accessing the Web for any sensitive information like e-mail or finance, always ensure that the site is using https.

**HTTPS Everywhere** [22] is a free extension downloadable for many popular browsers that ensures this is always in place. When creating your company website with **GoDaddy** [23] or a similar Web hosting service, it is important to include the Secure Sockets Layer (SSL) certification that supports https. It reassures customers and helps your ranking with search engines.

### 3.5.3. Web Tracking

All the major technology companies surveil Web usage for internal benefit, and/or sell the data to third-party data brokers and ad networks. Likely everyone has noticed something like this—mentioning an "Austin vacation" in an e-mail leads to travel ads around that city for days or weeks. The **Firefox** [24] browser offers a good compromise of security and usability for everyday business. **Privacy Badger** [25] is another free extension for popular browsers that automatically learns to block trackers of the kind running the travel ads.

There is no perfect combination of browsers and extensions in the cat-and-mouse game of Web privacy and surveillance. Even where Web-based ads and trackers are blocked, there is another subtler form of tracking called "fingerprinting" [26] that can still follow your interactions. So-called secure browsers such as **Tor** [27] are, however, unnecessary for everyday business and can be slow to use.

### 3.5.4. Bad Sites and Links

Just as with e-mails, texts, and calls, a website or link may not be what it says it is. Be especially wary when disclosing personal or financial information that you are, in fact, at https://BigBank. com and not http://BigBank.scam (also note the lack of "s" for secure once again). It is worth turning off previews where possible since they too can allow malicious content to be executed on your device.

## 3.6. **Phone**

### 3.6.1. Recommended Phones

**Apple**, **Pixel**, or **Samsung** are the smartphones of choice for startups and small businesses. These consumer-off-the-shelf (COTS) devices have a good mix of mobile hardware, friendly operating system (OS), app ecosystems, automatic updates, and encrypted backups. Generic brands tend to be weaker in supported lifetimes [28], and some even have malware [29] built in at the factory!

Source apps from the Apple or Google app stores, where they are current and vetted. Avoid "jailbreaking," "rooting," or "side-loading" apps, where hackers can use these venues to bypass security controls. Opt-out of location and data tracking, wherever possible. Some apps like **TikTok** are notorious for the large amount of personal information that they capture [30] and share overseas.

Use mobile monitoring tools like **Lookout** [31] or **Zimperium** [32] that add an extra layer of ongoing security and alerting. It was Lookout that first detected the Pegasus spyware [33] that has been used to target diplomats, journalists, politicians, and activists in the US and overseas.

## 3.6.2. Secure Phones "As Seen on TV"

Several years ago, there was a trend of custom-built "secure phones," the best known of which was the US Blackphone [34]. Trying to keep current with custom hardware, OS, and apps is difficult for such vendors, especially at low production runs and given the fast pace of mobile technology updates. These efforts typically translated into expensive and hard-to-use devices with short supported lifetimes. They are not recommended. Using something marketed as a "secure phone" with an unusual or dated appearance in its hardware or user interface of the kind seen in streaming dramas might leave you vulnerable with an effectively End of Life (EOL) device that attracts unwanted attention.

Larger enterprises manage fleets of standard (COTS) mobile devices with specialized tools like **MobileIron** [35] or **Workspace One** [36]. Where stronger security is needed in government, deployments might include custom ROMs (replacing the factory OS and updates) via specialist vendors like **CIS Mobile** [37] in their fleets, but the hardware, user interface, and apps are otherwise standard to retain usability and stay current.

## 3.6.3. Voice, Text, and Messaging

Calls and texts are the lifeblood of many small businesses and startups. Ninety-nine times out of 100, there should be no problem with their use. However, caller IDs are easily faked and both texts and calls can be intercepted by third parties even over the most recent 5G networks [38].

Secure end-to-end encrypted messaging, Voice Over Internet Protocol (VOIP) calls within a secure app, or video is better for sensitive communications. **Signal** [39] or **Wickr** [40] behind a trusted VPN are the recommended tools for sensitive communications. Even within a secure messaging tool, pay special attention that (a) E2E (end-to-end encryption) is enabled and (b) that every member of a group and its admins are part of your intended "circle of trust." Is "Uncle Vanya" really part of your investor team?

General video collaboration platforms like **Slack**, **Teams**, and **Zoom** [41] can be made secure with the right security configurations and subscription plans. However, their focus is mainly on ease of use given the tight timelines between calls for post-COVID workers. They should be treated as insecure, much like e-mails, texts, and calls. Special attention should be paid to messaging and file transfer within these tools since these may be archived and available for longer periods of time and to a much wider audience than intended.

### 3.6.4. Cars, Events, and Overseas Travel

Connecting your smartphone to a vehicle and/or other smart connected device may expose locations you have visited and other personal information to the vendor, their partner ecosystem, their advertising partners, and a myriad of third parties. It is particularly important to consider wiping information after connecting to a rental car. If traveling to industry events and/or traveling overseas, bring your own cables, chargers, and batteries and avoid leaving your devices unattended. An adversary agent can use the charging port to copy all your data and introduce malware that might not only track but also be used to impersonate you.

If traveling with sensitive data, use hardware-encrypted and PIN-enabled external storage like Apricorn [42]. If your personal device contains "the keys to the kingdom and the last few years of your IP work" then (a) check that you have current, secure, and verified backups ready for when it inevitably is broken or "lost" and (b) consider using a clean and near empty "burner" device with no biometric-enabled logins for events and overseas trips since adversaries may try to compromise your device physically or even subtly Over The Air (OTA) using baseband [43] or related attacks.

### 3.6.5. PACE Communications

The PACE acronym stands for Primary, Alternate, Contingency, and Emergency [44]. That means having a primary way of communicating, with several backups in case of outages and emergencies.

It is especially important when traveling. The Primary and Alternate can be easily met with cellphone service and locally with an Internet Service Provider (ISP) Wi-Fi. Wi-Fi pucks, SIMs, or e-SIMs from a different carrier are good options for a Contingency and when traveling. Satellite links as an emergency tool were historically expensive and limited but that is changing with newer options like SpaceX either standalone or built into newer cell phones [45]. Powering devices is also important, which can range from having a charged battery pack for a cellphone or laptop with you to generators and/or solar panels.

## 3.7. Post Quantum Cryptography

The US Government is introducing new National Institute of Standards and Technology (NIST) Post Quantum Cryptography (PQC) standards [46] from July 2022, to protect data at rest and in transit from cracking by adversaries using advanced, error-correcting quantum computers. The upgrade is a requirement [47] for US Government agencies. Today's encryption keys and algorithms are not expected to fail for another five to ten years [48]. The upgrade is beginning now since nation-state adversaries capture and store data now to decrypt it later (SNDL). This a problem if the data will remain sensitive in five to ten years time.

From a small business or startup perspective, this is something that is not addressable directly. It is, however, a desirable feature to look for when selecting secure communications providers, e.g., are they using quantum-hardened cryptographic keys [49] and/or supporting the new PQC algorithms [50]?

## 3.8. Conclusions

Security can be achieved affordably and easily through awareness, vigilance, and a thoughtful set of cloud-based tool choices and mainstream vendors for entrepreneurs and small businesses. Ease of use is a key consideration since it is the security you use that matters most. In all cases, your most important asset is not a high-technology tool but your own "Left of Bang" cybersecurity awareness, e.g.,

- Who is making and controlling your tools?

- Who is asking you for sensitive information from a website, a link, e-mail, text, or call?

- How are you staying current, e.g., updates, verified backups, and continuous monitoring?

- Are you encrypting data, limiting location and data sharing wherever you can?

- Are you taking special precautions when traveling to conferences and overseas and in case of outages and emergencies?

# References

1. https://amazon.com/Left-Bang-Marine-Combat-Program/dp/1936891301

2. https://www.forbes.com/sites/daveywinder/2022/10/30/former-uk-prime-minister-liz-trusss-phone-allegedly-hacked-by-kremlin-spies-report/?sh=2f151bf8308b

3. https://glacier.chat/

4. https://dashlane.com/

5. https://1password.com/

6. https://blog.mozilla.org/en/internet-culture/mozilla-explains/mozilla-explains-sim-swapping/

7. https://signal.org/en/

8. https://wickr.com/

9. https://lookout.com/

10. https://zimperium.com/

11. https://apricorn.com

12. https://signal.org/blog/cellebrite-vulnerabilities/

13. https://cisecurity.org/controls/cis-controls-list

14. https://www.cisa.gov/small-business

15. https://adint.cs.washington.edu/

16. https://inteltechniques.com/tools/index.html

17. https://baltimoresun.com/opinion/editorial/bs-ed-court-gps-20120125-story.html

18. https://cnbc.com/2022/07/01/google-will-delete-location-history-for-visits-to-abortion-clinics.html

19. https://openpgp.org

20. https://arxiv.org/abs/1510.08555

21. https://proton.me

22. https://www.eff.org/https-everywhere

23. https://www.godaddy.com/

24. https://www.mozilla.org

25. https://privacybadger.org

26. https://coveryourtracks.eff.org

27. https://www.torproject.org/

28. https://www.droid-life.com/2021/07/29/motorola-gives-up-on-long-term-updates/

29. https://kryptowire.com/news/kryptowire-discovers-mobile-phone-firmware-that-transmitted-personally-identifiable-information-pii-without-user-consent-or-disclosure/

30. https://www.cnet.com/news/tiktok-called-a-national-security-threat-heres-what-you-need-to-know/

31. https://lookout.com/

32. https://zimperium.com/

33. https://www.wsj.com/articles/firm-manipulated-iphone-software-to-allow-spying-report-says-1472149087

34. https://www.forbes.com/sites/thomasbrewster/2016/07/06/silent-circle-blackphone-losses-layoffs-geekphone-lawsuit/?sh=4e9f186e3b30

35. https://www.ivanti.com/products/ivanti-neurons-for-mdm

36. https://www.vmware.com/products/workspace-one.html

37. https://cismobile.com

38. https://ss8.com/resource/lawfulinterception_5g/

39. https://signal.org/en/

40. https://wickr.com/

41. https://gigazine.net/gsc_news/en/20200501-nsa-security-report/

42. https://apricorn.com

43. https://i.blackhat.com/us-18/Thu-August-9/us-18-Grassi-Exploitation-of-a-Modern-Smartphone-Baseband-wp.pdf

44. https://clintemerson.com/

45. https://www.washingtonpost.com/technology/2022/08/30/spacex-t-mobile-starlink-satellite/

46. https://www.nist.gov/news-events/news/2022/07/nist-announces-first-four-quantum-resistant-cryptographic-algorithms

47. https://www.nsa.gov/Press-Room/News-Highlights/Article/Article/3020175/president-biden-signs-memo-to-combat-quantum-computing-threat/

48. https://www.telegraph.co.uk/technology/2020/01/22/googles-sundar-pichai-quantum-computing-could-end-encryption/

49. https://www.quantinuum.com/products/cybersecurity

50. https://www.nextgov.com/ideas/2022/07/understanding-nists-post-quantum-encryption-standardization-and-next-steps-federal-cisos/374792/

# 4

# Protect Your Financial Transactions Now! Cybersecurity and Finance for the Entrepreneur

Chris Sundberg

## 4.1. Introduction

As an entrepreneur, you have many of the basics in place to execute your business plan. This chapter will focus on what to look for in financial systems from a cybersecurity point of view. Not too long ago, when an entrepreneur developed a business plan and needed to execute it, they would go to a local bank and have an in-person meeting with a representative and talk about the range of options available from the bank to the entrepreneur to help make the venture successful.

How does the cash get from point "a" to point "b" in a business if you are an entrepreneur? How much thought has gone into your business plan with respect to how financial services are used? How do you know if your financial services are secure?

## 4.2.  A Little Background on Data Breaches that an Entrepreneur Should Consider

The IBM "Cost of a Data Breach Report 2022" documented the cost of data breaches across 17 industries looking at mitigations like incident response (IR), zero-trust, end-point security, and artificial intelligence technologies. For example, the average cost of a data breach is 5.92 million USD in an organization without an IR plan. A mature IR strategy will cut that cost by about 50%. The study, conducted annually, has found that most organizations have experienced more than one data breach [1]. Even with repeat attacks against organizations, learning from past events helps secure services that someone may use. Commonly accepted statistics indicate that 75% of breaches involve hacking and malware, 18% are due to accidental disclosure, 6% involve insider threats, and 2% are physical breaches [2].

A good security strategy plan to protect and store financial information should incorporate what data could be stored and trusted with an external application and what type of data should be backed up and stored locally with local access. Financial data governance and controls around who is authorized to spend or approve money are pieces of an overall security framework.

Attackers today like to take advantage of the anonymity of Internet communication to entice otherwise innocent workers into disclosing or disbursing money. These tactics look like:

- Business e-mail compromise (BEC): Criminals send an e-mail message that appears to come from a known source making a legitimate request.

- Phishing: A target or targets are contacted by e-mail, telephone, or text message by someone posing as a legitimate institution to lure individuals into providing sensitive data.

- Stolen or compromised credentials: Specific data or authentication tools are used to identify the user, authenticate users, and grant access. Stolen credentials put the identity and authentication information in the hands of an attacker to achieve system entry.

**FIGURE 4.1**    Peter uses open Wi-Fi at the coffee shop.

Illustrated by Phillip Wandyez.

*Peter was not much of a morning person, but he enjoyed going to his local café for a coffee and to check up on his morning work tasks. "I sure am glad that my coffee shop offers free Wi-Fi," Peter thought to himself as he checked his personal and business emails and his business bank account. Unfortunately for Peter, the mysterious person sitting in the corner of the coffee shop in a hoodie was actually a hacker who was quietly monitoring the café's Wi-Fi traffic and was able to obtain Peter's user ID and the passwords to his email and bank account.*

# 4.3. How to Keep Your Finances Safe

Policies and procedures around transactions help mitigate the risk of these tactics being successful. Awareness training in recognizing an attempt at BEC or phishing plays an important part in any information security strategy. Clear definitions should be in place regarding how and why an authorization to spend may occur. Firm rules or multistep approval processes to approve the disbursement or disclosure of information is an important technique to establish governance rules within an organization.

Security controls with respect to the storage and use of data include hardening the system that accesses the data as well as a sound backup strategy to provide for business continuity. Financial data are the lifeblood of any company. Data can be accessed and stored locally or in the cloud. Often the financial information will include bank accounts and information about business relationships and may also include a category of information known as personally identifiable information (PII). PII is the subject of privacy and data protection laws. There are specific rules often regarding this type of information. This chapter, however, will concentrate on financial information topics and not privacy.

## 4.3.1. Identity and Access Management (IAM)

Multi-Factor Authentication (MFA) is a tactic that can be implemented regardless of whether the platform is local (on premise) or in the cloud. MFA should be used to replace the classic username/password (single factor) authentication. MFA is a tool providing greater granularity of:

- Where a login may originate.
- Who is making the request?
- Is there a measure of legitimacy to the request?

Operating systems like macOS, Windows, and Linux and Web-based applications commonly support MFA with multiple ways to enable the authentication (SMS or text messaging, hard tokens like Titan Key and Yubico FIDO, and authentication applications on phones).

## 4.3.2. Data Encryption

Data encryption provides a way to ensure confidentiality and integrity of data. Encryption can cover data in flight or transmission. Encryption also covers data at rest or storage. Data transmission should be evaluated for what is contained in the data being sent:

- Is the data sensitive (account number, financial transactions, PII)?
- Does the data require integrity?
- Is the recipient trusted?

These factors will dictate the amount of encryption that may be required to communicate. Usually, this is handled today through modern, secure protocols. Protocols such as the latest version of Transport Layer Security, or TLS, are typically invisible to the end user. However, even though the latest version of TLS as of this writing is 1.3, vulnerabilities have been found in prior versions which have necessitated updates.

At rest, encryption is offered standard as part of operating systems, such as Microsoft's BitLocker or FileVault2 on macOS as well as Linux and cloud-based storage. Contemporary hardware available to an entrepreneur should have encryption technology support out of the box. Check for the presence of a Trusted Protection Module on the system. Disk-level encryption protects the data from being observed by someone who is not authorized if the storage medium, like a hard drive, computer, or USB drive, is stolen.

## 4.3.3. Business Continuity (Backup and Restore!)

Beyond just encrypting the data, a sound backup and restore strategy need to be part of the overall security strategy plan. In addition to design and intellectual property, financial data pertinent to the entrepreneur should be part of a backup strategy.

These backups should follow common best practices such as a 3-2-1 backup strategy:

- Three copies of your data: this includes your original data and two duplicate versions in case one of your backup options becomes corrupted, lost, or stolen.

- Two storage types being used in the event a failed backup or recovery is due to the specific type of storage option.

- One copy is stored away from your home or business in case there is a disaster that damages or destroys the property.

The same 3-2-1 strategy that an entrepreneur may incorporate at the local level should be evaluated on the cloud provider level if significant data are stored on the cloud. Cloud and application providers typically provide their strategies or offer service-level agreements around backup and data availability.

This might seem like quite a bit of protection around financial data. Many well-established businesses still struggle with the basic concepts of data security in an overall business strategy. As such an entrepreneur has a unique opportunity to put best practices into action early in a company's lifecycle. Managed service providers (MSP) can provide guidance in many security discussions around financial data.

## 4.4. The Cloud, Data, and Software-as-a-Service

Technology has created many financial service options for the entrepreneur that have not been available in the past. Easy access to many more banking options, loans, payment services, and investment opportunities has given the modern entrepreneur many more tools in the toolbox than in past generations of entrepreneurs. This growth of services coincides with the growth of applications and platforms leveraging the power of the Internet and the cloud.

Online tools based on the "cloud" will use a technology known as "Software-as-a-Service," or SaaS. This revolution in computing enables companies to offer a wide range of applications to customers with new features developed and deployed in a rapid amount of time. As an Internet-based offering, the end user will use a Web browser to interact with an application. On the back end, these applications can be hosted on one or more providers, providing resilience in case one of the systems is "down."

This technology underpins a wide variety of applications we use today. Financial services are no exception and are typically held to a higher standard from a security and assurance point of view. However, financial services and applications are exposed to many of the same risks, such as ransomware attacks, hackers, poor security configurations, and insider threats, that many SaaS applications face.

There are a number of tools that can help determine the security posture of a cloud service. The American Institute of Certified Public Accountants (AICPA) provides a method of reporting called system and organization controls (SOC). SOC has a variety of levels that can be audited against. SOC controls are designed to build trust and assurance in services provided by cloud vendors to meet specific user needs [3]. There are three types of SOC reports:

- SOC 1 for Service Organizations: Internal Controls over Financial Reporting (ICFR).
- SOC 2 for Service Organizations: Trust Services Criteria.
- SOC 3 for Service Organizations: Trust Services Criteria for General Use Report.

Evaluating cloud data storage or SaaS applications relies on SOC reports. These reports should be available from a provider upon request and used as a tool to evaluate a platform's risk that may host financial applications the entrepreneur may need to access. For example, a SOC 2+ report is usually sufficient for evaluating security and trust assurance of data and applications around the platform. The SOC 2 framework contains a comprehensive set of criteria known as the Trust Services Principles composed of five sections:

1. The security of a service organization's system.
2. The availability of a service organization's system.
3. The processing integrity of a service organization's system.
4. The confidentiality of the information that the service organization's system processes or maintains for user entities.
5. The privacy of personal information that the service organization collects, uses, retains, discloses, and disposes of for user entities.

The cloud and SaaS provide a type of force multiplier for the entrepreneur that has not been present before. With this rich set of data, the entrepreneur needs to examine the data protection around the application that shall be used.

An "insider threat" may not be inside the company. An example is the 2019 Capital One data breach involving a software engineer at a cloud provider. The engineer was able to capture credentials allowing privilege escalation and access to data on over 100 million people including Social Security numbers and bank account numbers. This is a topic to talk to the SaaS provider about depending on the importance of the data stored on the cloud.

### 4.4.1. API Security

The ability to bring in richer sets of data to gain visibility into the success of an entrepreneur's business plan is enabled by interfaces into different applications based on SaaS. Often pursuing additional funding, a startup will collect data on key performance indicators (KPIs) of interest to investors. These data in the past had been collected through manual means of local-based applications, spreadsheets, and data. Automation of these data can provide faster turnaround times for collecting and reporting on KPIs that interest investors. These data are often stored in a mix of applications such as accounting software, payment management software, logistics management, and payroll management. SaaS can provide an integration of these items into a "dashboard" using Application Programming Interfaces (APIs).

The API can be used to access data, and there are three types of APIs to keep in mind [4]:

1. Public (aka Open) APIs are most common. They are provided by companies to allow anyone to connect to their services.

2. Partner APIs are those shared between contracted business partners and not available publicly.

3. Private (aka Internal) APIs are used only within a company.

Bringing about the use of an API to gather financial data requires thought about where the data sources are accessed, how they are accessed, and the involvement of stakeholders to enable the gathering of information. A business plan for the entrepreneur should outline when this type of automation makes sense in the scale-up of the business. API is a very powerful tool to secure this type of information. Especially when financial data are concerned, one should look at specific security standards such as the OWASP Application Security Verification Standard. These help security assurance in the use of API.

## 4.5. Credit Card Processing Compliance and Standards

Almost every company needs to handle payments. A bit of a primer with respect to payment is helpful to discuss credit card processing in this section. ACH, or Automated Clearing House, is a network that directly transfers funds from one bank account to another. ACH fees are typically very low, but the processing time for an ACH transaction may take as long as five days, while a credit card payment can process between one and two days [5].

Credit card transactions are used as a method of payment by consumers. The merchant's bank is sent details of the transaction, forwarding the transaction to a credit card network such as Visa or Mastercard. The credit card network clears the payment and then seeks authorization from the issuing bank (the customer's bank). An authorization phase follows next where the issuing bank analyzes details like Card Verification Value (CVV) code and the transaction to be vetted as legitimate. Finally, clearing and settling the transaction occurs (unless the customer contests the payment) [5].

Often, a business may opt for a credit card processing solution like Square, Clover, or PayPal to abstract the details of handling and processing credit card transactions. The solutions employ an off-the-shelf tablet and connected reader. An application on the tablet securely connects to the reader and transmits credit card transactions to the provider. The solution provider companies usually charge a small percentage fee on a transaction to cover the costs of credit card processing.

If the business will be processing or using credit cards, there is a strong set of compliance standards a business or SaaS service must follow. This standard is the Payment Card Industry (PCI) Data Security Standard (DSS). This is a global information security standard designed to prevent fraud through increased control of credit card data. PCI DSS is widely adopted in industry but is not required by laws from agencies or governments. If a vendor or service does process payment card data, they will more than likely have a report on their PCI DSS compliance through a Help or About section on their website.

In addition to PCI DSS, many financial companies an entrepreneur will do business with must comply with other regulatory standards. Because of the amount of financial gain to be made by a successful attack on a financial target, the level of cybersecurity regulation in this field is especially strong. Some recent examples [6]:

- New York State Department of Financial Services Cybersecurity Requirements Regulation for Financial Services Companies Part 500 (NY CRR 500) of Title 23.

- US Securities and Exchange Commission (SEC) interpretive cybersecurity guidance.

- National Cybersecurity Center of Excellence (NCCoE) NIST Cybersecurity practice guides (SP 1800-5, SP 1800-9, and SP 1800-18.

- Twenty-four states have passed bills or resolutions related to cybersecurity.

Incorporating regulation in the financial area helps address concerns with security that customers or users may have. While the burden for security is high, it is commensurate with the amount of reward an attacker may receive if there is a successful breach of service.

## 4.6. Conclusions

Financial cybersecurity is a complex and hard topic. This chapter covered some fundamentals to help protect money and provide business continuity. Some of these best practices include:

- Protect financial systems using strong authentication techniques like MFA.
- Develop and maintain business continuity strategies like backup and restore.
- Perform due diligence with cloud and SaaS vendors; Security of API.
- Follow secure money processing strategies.

## References

1. IBM Security, "Cost of a Data Breach Report 2022," IBM Corporation, Armonk, NY, 2022.

2. Cybersecurity Guide, "Cybersecurity in the Financial Services Industry," 2022, accessed September 6, 2022, https://cybersecurityguide.org/industries/financial/.

3. American Institute of Certified Public Accountants (AICPA), "SOC for Service Organizations: Information for Service Organizations," 2021, accessed September 6, 2022, https://us.aicpa.org/interestareas/frc/assuranceadvisoryservices/serviceorganization-smanagement.

4. Moore, R., "API Security Best Practices for the Financial Industry," IEEE Computer Society, 2022, accessed September 6, 2022, https://www.computer.org/publications/tech-news/build-your-career/api-security-best-practices-for-the-financial-industry.

5. Rotessa, "ACH vs Credit Cards," 2022, accessed September 6, 2022, https://rotessa.com/resources/ach/ach-vs-credit-cards/.

6. Harvey, S., "Top 4 Cybersecurity Challenges Facing the Financial Services Industry," Kirkpatrick Price, 2019, accessed September 6, 2022, https://kirkpatrickprice.com/blog/top-4-cybersecurity-challenges-facing-the-financial-services-industry/.

# 5

# Who Needs a VPN?

Dennis Vadura

## 5.1. Introduction

Do I need a VPN? The answer to the question depends on who you are and who you ask. In many instances, the people offering the answer do not fully understand what a VPN is and what it can and cannot do. This chapter is a commonsense approach to the answer. But before we answer the question we should spend a few words to, in simple language, teach what a VPN is.

## 5.2. What Is a VPN?

VPN is an acronym for ***Virtual Private Network***. There, now you know everything you should know about VPNs. Well… um…, okay, let us break it down. ***Network*** is the easy term, it just means any infrastructure that provides a connected path between two devices thereby enabling them to exchange information. ***Private***, we think: "That should be obvious right?" The intuitive meaning is something that is private is impervious to prying eyes, but it could also mean personalized, as in, it is only for me.

Of course, it could mean both. Therefore, in this document, we define Private to mean both impervious to inspection and personalized. *Virtual* is the term that most people struggle with. This is because it is not a common practice to interact with virtual things. Your spouse is not virtual, your car is not virtual, your cat is not virtual, etc. Most people do not have a good feel for what virtual is. That is unless you are an information technology (IT) professional.

As used here, virtual means that there is no physical corresponding network wire that identifies the connection. It is an abstract connection that utilizes actual real-world wired or wireless connections to work. It is similar in concept to online stores that have no physical storefront. The store exists only as long as the metadata that defines the storefront exists, is current, and is available to online users. In the context of a virtual connection, the connection exists as long as the parameters that describe the connection remain valid and the connection is available to be used by a user.

So what exactly is a VPN? VPNs started out as the Point-to-Point Tunneling Protocol (PPTP) developed by Microsoft for implementing Point-To-Point network connections. These were designed as connections that ran on top of (were virtual) existing real-world connections and allowed two computers on either end of the PPTP connection to communicate even though they were not directly connected and not directly addressable via the local or wide area networks in use at the time. To achieve this, the PPTP established a tunnel. The tunnel ran virtually from one network to the other and had four key properties.

1. A local address on each end that was addressable and reachable by computers connected to the network at that location.

2. Any information sent to the local address on one end would be delivered to the other end of the tunnel and passed on to the destination computer on the other network.

3. To accomplish (2), the tunnel PPTP implementation had to take the original packet being sent and wrap it in a bearer packet, just like putting a letter into an envelope that has as its source and destination the two tunnel endpoint addresses.

4. The final piece of the puzzle was an appropriate set of routes that needed to be added to local routing tables so that computers on one end knew that computers at the other end were reachable through a local tunnel address as the gateway.

When configured correctly, traffic between two PPTP connected computers would work as if they were sitting next to each other and were connected to the same local network. What made this interesting is that you could set this up at the office level or all the way down to two individual computers. This is essentially a VPN, that is, *virtual* and *networked* but is *not private*. PPTP became a VPN the minute that the tunneled traffic was encrypted. Modern VPN systems all operate the same way. They all provide tunnels between networked devices or systems, they typically encrypt the wrapped packet payloads, and they add routes to capture some or all traffic flows on tunnel creation. They can be operated by individuals for their own personal benefit, by entities that provide a public VPN service to individuals, or by corporations for the benefit of their employees.

There that is it. Now you are a VPN expert. In principle, to implement a VPN, you take a PPTP tunnel, mix in a little cryptography, and voila you have a VPN. Anyone can do it. However, as with most things, the devil is in the details. Trusted VPN solutions offer good privacy, good performance, limited logging, and are simple to use.

The remainder of this chapter will focus on why you should use a VPN, and if you do, what properties of a VPN you should look for when selecting one. This is certainly not going to be a developer's guide to creating a VPN, but it will get somewhat technical. Thus, you have been warned.

Illustrated by Phillip Wandyez.

**FIGURE 5.1** Peter's customer data gets compromised.

*Peter was not just a random mark. He was a highly desirable target because Peter was secretly rich. Peter showed no obvious signs of his wealth, in fact, he made a point of being just like everyone else. Peter had been a lifer, he started during the early days of Casco and spent 30 years rising to Executive VP of Sales. For those with a keen eye for detail, the signs that Peter was well off were subtle but visible. He lived in the exclusive part of town in a meticulously landscaped house. He wore an expensive watch and while he did not drive a luxury vehicle, he did come to the coffee shop every morning and stayed until noon. He became a class-A target when his bank account balance was verified by the hooded hacker that stole his credentials using the public Wi-Fi network in the coffee shop. It did not have to be that way.*

# 5.3. **But No One Is Spying on Me**

Why should anyone spend money to use a VPN service in the first place? The gist of a common argument often made by those discouraging the use of VPN services goes like this:

"The current network technologies that are used for daily browsing are secure enough and no one can get enough information about me to be useful. HTTPS is secure, my IP is shared by others, and I have told my browser to not share my location unless I explicitly allow it. The Top 100,000 websites all use the HTTPS protocol to encrypt traffic."

HTTPS stands for "Hypertext Transfer Protocol Secure." It is the protocol that is used between your browser and an Internet server and is used to securely (i.e., encrypted) deliver webpage information to your browser for your viewing pleasure. HTTPS does not specify the encryption; rather, it relies on Transport Layer Security (TLS) to encrypt the data packets between a client and a server. Think of HTTPS as specifying the language to speak and TLS specifies the encoding of that language for transmission purposes.

The statements in the preceding quote are true. To a reasonable person, it would seem that provided you use a modern Web browser, your information is secure and no further action need be taken. In fact, proponents of "no VPN is needed" explain this and concede that a VPN is useful only if you are an employee who either travels a lot or works remotely. In those cases, a VPN together with a device trust policy and possibly an authentication token makes sense when accessing your employer's internal network resources. Proper business use of a VPN is a whole separate topic unto itself.

The core of the argument revolves around HTTPS and the assertion that it is secure. This is largely true. But it is true only if the servers that you are visiting are exclusively using HTTPS, have configured their HTTPS to not use deprecated or outdated Secure Sockets Layer (SSL) or TLS algorithms, and use a trusted certificate authority to issue their TLS certificates. As recommendations evolve, using the right algorithms requires vigilance by the Web service operator. For example, SSLV2, SSLV3, and TLS 1.0 are no longer supported by many servers, and the use of TLS 1.1 is discouraged in favor of TLS 1.2 and TLS 1.3. TLS 1.3 (RFC 8446) is the latest version of the TLS standard and provides enhanced security and faster HTTPS session setup.

In TLS 1.3, enhanced security is provided by not supporting cipher suites that have known vulnerabilities and rejection of all algorithms not supporting perfect forward secrecy. To accomplish the latter, all static key exchanges are disallowed as is RSA [1]. This leaves ephemeral Diffie-Hellman [2] key exchange as the only supported key exchange protocol and allows for faster handshake negotiations as the key exchange protocol is known by definition rather than negotiation.

The question to ask is how many of the 100,000+ websites using HTTPS have in-house or contracted IT support that has kept up with the above details? The answer to the question is important because the strong security of HTTPS is the core argument that is used by the "no VPN is Needed" advocates.

TLS 1.3 was created explicitly to handle the issues surrounding compromised and static key exchange protocols. Any website not using TLS 1.3 to secure HTTPS is thus possibly hackable by a determined adversary using known vulnerabilities, and once compromised, your privacy as it relates to that website or service is also compromised. It is highly likely that most of the 100,000+ websites are not following current best practices as they apply to TLS 1.3 and HTTPS. It is clear from the discussion here that it is a complex issue requiring the right expertise. Even if you get the TLS and HTTPS configuration perfectly correct, there is little that can be done with a compromised certificate authority.

Websites or services that use HTTPS are vulnerable to Certificate Authority attacks. StartCom is a well-known case of a compromised certificate authority. StartCom stopped issuing new certificates on January 1, 2018. The action shut the company down. The shutdown occurred after all major browser vendors removed StartCom root certificates from their browsers. The browser vendors took unprecedented action after it became clear that StartCom was secretly acquired by WoSign Limited [3] (Shenzhen, Guangdong, China) through multiple intermediary companies. A fact that was only revealed by a Mozilla investigation related to another root certificate issue. What was at stake was simply the fact that every StartCom certificate ever issued was now compromised, and if used to secure HTTPS handshakes, then the resulting sessions could be decrypted by a nefarious third party.

If a website or its servers are compromised by a man-in-the-middle certificate authority attack, then your activity with the service provided by the website is also compromised. The easiest network location to effect a successful man-in-the-middle attack is at the edge routers that you interact with (e.g., at a hotel) or in Wi-Fi access points or at your Internet service provider (ISP), with the ISP being more difficult to compromise than the former.

In the case of a compromised certificate authority, the use of a VPN is useful because it can reroute your traffic so that a clandestine man-in-the-middle at your ISP or at an edge router is bypassed. This is particularly true if your VPN endpoint is external to the jurisdiction in which the compromised router or access point is located.

For the remainder of this discussion, the assumption is that all of the websites you visit are configured to the latest standards and are using certificates and keys issued by reputable certificate authorities (e.g., Let's Encrypt). We have thus moved the threat envelope away from website and man-in-the-middle compromises (where a VPN provides limited value) to traffic-based attacks where a VPN is the only way to provide protection beyond that which is provided by the TLS.

Traffic-based attacks rely on collecting packets and storing them (so-called netflow data). Netflow data are captured at the ISP level and sold to private companies that then coalesce, tag, and provide the data to anyone able to pay the requested price. At a high level, netflow data create a picture of traffic flow and volume across a network. It can show which server communicated with another, information that may ordinarily only be available to the server owner or the ISP carrying the traffic. Crucially, these data can be used to, among other things, identify patterns of use for individual users. When coupled with an HTTPS certificate compromise or TLS algorithm compromise this results in an attacker's ability to obtain clear text (the unencrypted original text) from HTTPS, provided all of the packets are available for post analysis. The clear text then provides detailed information about websites being visited, user location, account information, password information, etc. All of which allow for further compromise of related systems and accounts.

Thanks to the European Union General Data Protection Regulation (GDPR) every webpage using tracking cookies displays a dialog warning us that the website uses tracking cookies; as a result, we blindly click "okay" to make the dialog disappear. The result is that cookie information can still be used to track a user's activity across many different websites and services. The cookies are used to help provide the Internet protocol (IP) address to geolocation mappings that allow third parties to know your approximate location (to within 10 m). This can be achieved even if you are using a dynamically assigned Internet address provided by your ISP.

If you are an individual that values privacy, then taking steps to render the analysis of netflow data less useful may be the only way to provide you with both location and information privacy. A VPN is the only way to do that.

This is all happening because profile building is useful in big business and big government. Both businesses and governments have profiled their customers and citizens. Mining the profile data then results in potential actionable information for business purposes or government oversight or, worse, government-driven influence over your constitutionally protected freedoms.

For example, Google wants to know who is interested in what—critical for ad serving; Meta wants to know who you are and what you like—critical for ad serving. Both want to know what you buy and when you are likely to buy again—critical for ad serving. Conversion and influence of your buying decisions depends on habit intelligence. The value of your ad to an advertiser is tied to your purchasing power.

If you are someone with a 6+ digit bank balance, then you become even more interesting. Governments want to know who your friends are, where you work, which co-workers you hang out with, who you party with, what your political beliefs are, what your religious beliefs are, whether you are married, whether you have children, where you live, where you take a vacation, and so on. Don't believe me? Recently it was made known that information on all election poll workers in Michigan was stored on servers in China and included information like their birth date, whether they have children, and what schools those children attend. Information that "could" be used to influence your election night duties.

It should by now be clear that the desire to know everything about your online and offline life is of interest to many parties. Your electronic DNA is very interesting. So let us ask the question again.

## 5.4.  Do I Need to Use a VPN When Surfing the Internet?

In short yes, and if you think that your life is known already and that you are uninteresting then consider some of these:

- *No one cares if I have cancer, AIDS, or COVID.* A potential employer might, an insurance company will certainly want to know when offering you a policy price.

- *No one cares that I visit with my poker buddies on Friday night and go to the Elks or Freemason lodge on Saturday night.* A government agency looking into your political affiliation might.

- *No one cares if I work at Google, Microsoft, or the CIA.* Except for the foreign agent that is looking for someone to solicit information.

- *No one cares if I drop my child off at daycare.* Except for someone who is looking for leverage against you.

In summary, anyone looking for actionable or financial leverage against you desires to know all of the above things and more.

## 5.4.1. Wow! It Is Hopeless!

OMG, everyone is spying on me. True, they are. It is not hopeless though. You should consider the secure envelope model. Everyone can see the from and to address on the envelope but the contents are unknown. This is in fact what a good VPN does. It makes the contents unknown. A great VPN makes it relatively hard for netflow providers to consolidate individual flows from many sources. For this reason, VPNs are under attack because they work. You just have to find a good one.

## 5.4.2. Threats to VPN Traffic Are Everywhere!

Since the mass adoption of remote work, VPN threats have increased by 2000% [4]. In the first quarter of 2021, there was a 1,916% increase in attacks against Fortinet's SSL-VPN and a 1,527% increase in Pulse Connect Secure VPN. Finding VPN vulnerabilities allows a threat actor to gain access to a network. Data recording for replay attacks is a popular strategy if HTTPS traffic can be captured. To capture the traffic an adversary will compromise a Wi-Fi access point, a router, or any other piece of interesting networking equipment that is in the data path of interest. The captured data can be used for offline man-in-the-middle attacks.

Other forms of attack include random number generator attacks that can be used to compromise key generation and selection, packet length, and packet timing attacks that can be used to infer meaning from the length of transmitted packets.

Internal threats are also an issue. Notably, NordVPN was hacked back in the first quarter of 2018 and SSL encryption keys were compromised [5]. Whether these could further be used to compromise traffic is unclear. It would have to be a careless service deployment that would result in the use of the website keys for securing the VPN API and actual data traffic. A similar set of circumstances hit TorGuard and VikingVPN and compromised a squid proxy certificate used by TorGuard.

## 5.5. What Features Matter Most in a Modern VPN Service?

What are the important features of a modern VPN service? Here are some key items to consider. A VPN service that adheres to these principles is well on its way to helping keep you and your data secure.

1. You get what you pay for. VPN services cost money to run. Only use a reputable service that charges a reasonable fee. If it seems too low or is a lifetime purchase then they are selling your data! You have been warned!

2. Use a service headquartered in a reasonable legal jurisdiction. All jurisdictions can force any business to provide them information on a customer. So pick a VPN domiciled in a jurisdiction where at least you have a chance to know that you have been targeted.

3. Use a VPN that optionally detects and restricts trackers.

4. Use a VPN that keeps minimal session logs and no traffic logs.

5. Use a VPN that routes your client traffic through multiple servers while connected (not just one server), thereby making netflow collection more difficult.

6. Use a VPN that bonds multiple physical network paths into a single tunnel—this makes it harder to capture all netflow data for your network sessions.

7. Use a VPN that hides in plain sight in the middle of other traffic.

8. Use a VPN that has perfect forward secrecy with frequent key rotation.

9. Use a VPN that tells you if data from your device or computer are going to known threat actors—notifies data exfiltration.

10. Use a VPN that uses the latest security suites.

11. Use a VPN that does not rely on an external third party for the issuance of certificates (i.e., the VPN runs its own internal Certificate Authority).
12. Use a VPN that provides packet length obfuscation.
13. Use a VPN that can steer sessions to multiple network exit points.

There are other more technical attributes, but the above list is sufficient to weed out the VPNs that want to steal and sell your data, that allow data to be exfiltrated to parties that should not have access to it, and that have poor or nonexistent security policies.

Keep in mind that there is a large body of VPN implementations not discussed here. Here we focused on the public consumer VPN options. We purposely left out discussion of network layer-1 and layer-2 based VPNs because they rely on different routing and packet wrapping technologies and would unnecessarily complicate this discussion.

That being said, you should, after reading this, be empowered to look critically at the important feature set of a VPN irrespective of the network layer in which it operates. The security considerations discussed here apply to a VPN operating in any Open Systems Interconnection (OSI) network layer. Furthermore, a good VPN will help you limit tracking your location and will provide you the ability to jump jurisdiction so that you can be relatively certain that your traffic is not compromised by a local threat actor.

## 5.6. **It Is 2023, What VPN Fits the Bill?**

We just covered a lot of details. To help you choose a VPN, use the following figure to choose a privacy-focused VPN for your needs (Figure 5.2). This figure is good for 2023, if you read this in years beyond 2023 then remember, free or lifetime subscribed VPNs are a scam, VPNs that do not help to obfuscate netflow collection pose a security risk, and VPNs that do not provide adequate region hopping may not protect you from a local man-in-the-middle threat actor. If you are a typical VPN user and use a VPN on your laptop or mobile device, then this figure is for you; if you are a corporate IT security professional, then this figure provides a summary of the feature set you should look for from your corporate VPN provider.

**FIGURE 5.2**　Important VPN features and which VPNs provide them.

| VPN Service | Latest Security Suites | Frequent Key Rotation | Notifies Data Exfiltration | Single VPN Session via Multiple Servers | Packet Length Hiding | Region Hopping | No Packet Logs | Minimal Session Logs |
|---|---|---|---|---|---|---|---|---|
| Atlas VPN | ✔ | ✔ | | | ✔ | ✔ | ✔ | ✔ |
| Cyber Ghost VPN | ✔ | | | | | ✔ | ✔ | ✔ |
| Express VPN | ✔ | ✔ | | | | ✔ | ✔ | ✔ |
| Feather VPN | ✔ | ✔ | ✔ | ✔ | ✔ | ✔ | ✔ | ✔ |
| Nord VPN | ✔ | | | | | ✔ | ✔ | ✔ |
| Proton VPN | ✔ | ✔ | | | | ✔ | ✔ | ✔ |
| Tunnel Bear | ✔ | | | | | ✔ | ✔ | ✔ |

Figure 5.2 lists features unlike those most often cited by VPN comparison shopping sites. This is because items like the number of servers and countries for those servers and whether video streaming is supported or not provide no value to the most important VPN feature: Is your privacy protected? For this reason, FeatherVPN stands out. It not only provides the latest cryptographic ciphers, but it is also architected to provide the best possible protection against replay and data-flow gathering attacks.

## 5.7. Conclusions

It is 2023 you are the most important product for many organizations and startups. Gone are the *total accessible market* and the *flywheel effect*. The new king in town is "lifetime customer value."

The change reflects the harsh reality that customer acquisition is costly and, in the short term, may not be covered by the service revenue generated from the customer. Cultivating a long-term customer relationship is key to maximizing "lifetime customer value." Creating a profile about the customer is key to the success of these newish side channel monetization strategies.

The key aspects of collecting a customer profile are:

1. Precise current and historical location.
2. Private and commercial affiliations.
3. Your purchasing habits including time, product, and seller.
4. Your circle of friends, co-workers, and loose social relationships.
5. Who you message, when you message, and what you message.
6. Your browsing history including time, date, keywords, and URLs.

All of these are immensely valuable when combined into a single profile that can be mined with a variety of queries whose results can be used for both benign and nefarious purposes. Two notable examples are TikTok—which is being used to profile and track billions of users and whose profile is available to authorities in China. India has banned TikTok from the country, and other governments are considering doing the same. The second is the push to add "smart" features to televisions and other consumer devices. By adding intelligence to televisions (TVs), TV manufacturers have realized that the long-term value of a TV is not the initial sale but the resulting data mining that it affords. This is a new revenue stream for many that did not exist before. To point out, data tracking companies such as Inscape [6] and Samba [7] openly and proudly discuss the TV manufacturers that they partner with.

For the TV manufacturers, it is "post-purchase monetization" or in the language of Silicon Valley "lifetime customer value." History has taught us that too much of a good thing is never a good thing. Companies such as Inscape and Samba exist to aggregate disparate profile information into a single customer/user profile and then monetize access to that treasure trove of data. A VPN can help blunt the stream that flows to these third-party profile aggregators.

# References

1. http://www.internet-computer-security.com/VPN-Guide/RSA.html

2. https://www.rfc-editor.org/rfc/rfc5246—Section 7.2 defines the details of the exchange

3. https://en.wikipedia.org/wiki/WoSign

4. https://www.darkreading.com/attacks-breaches/vpn-attacks-surged-in-first-quarter/d/d-id/1341300

5. https://www.techradar.com/news/whats-the-truth-about-the-nordvpn-breach-heres-what-we-now-know

6. https://www.inscape.tv

7. https://www.samba.tv

# 6

# Securing Your IoT Devices

Peter Laitin

## 6.1. Introduction

IoT is a relatively new buzzword that affects all of us in some way, shape, or form, and understanding its origins is important. The Internet of Things (IoT) has been around since DARPA (Defense Advanced Research Projects Agency) began working on it in the 1960s, but only recently has it begun to take off. The Internet was exposed to commercial usage in the early 1980s when several carriers began selling dial-up access services. We are seeing even more growth with the advent of faster networks, processors, and cheaper devices that can communicate remotely via Wi-Fi or Bluetooth (or both).

Folklore says that Carnegie Mellon students in the early 1980s hooked up a Coke vending machine to the internal network to tell when the machine was out of drinks, so they did not waste time shuffling across campus only to find it was empty.

IoT was created to accelerate the exchange of information electronically between two or many parties, and today all aspects of business and personal life leverage it for the same reason. But for every convenience, there is always a lousy actor wanting to exploit it. IoT is no exception, so security is critical to protect the most important data of any entrepreneur.

Entrepreneurs come in many shapes and sizes and different verticals, so briefly discussing each will show the commonalities and differences in whatever business you latch onto. The ongoing need for communication drives the constant use of cell phones, computers, laptops, tablets, and just about any device that connects to the Internet to transmit and exchange data. The invention of technologies to automate many manual processes has become a great saver of time and resources, depending on your field of business, so briefly touching on each will cast clarity on the need for security for IoT.

Addressing the basic IoT needs of any entrepreneur is a great place to start, and understanding their value and vulnerabilities is essential. We will discuss the anecdotes to have a healthy IoT experience.

# 6.2. Reduce Your Attack Surface

Services offered by an IoT device should only be accessible by the owner and the people in their immediate environment whom they trust. However, this needs to be enforced more by the security system of the IoT device. Any device that can access services offered by an IoT device can potentially access its functionality. Attackers use this fact to initiate attacks on the system to alter the behavior or take complete control of the device. Because many of today's IoT devices do not enforce adequate access control and authentication, they often expose their functionalities to unauthorized parties. This can cause an attacker to either steal private data (e.g., photos) or even remotely perform actions on a device that an attacker should not be able to do (e.g., turning off your alarm system remotely).

It is a common security strategy to attempt to reduce the attack surface. An attack surface is the sum of vulnerabilities, pathways, or methods that hackers can use to gain unauthorized access to your network or sensitive data. It is one of the first steps in securing a system. There are many ways a device can be attacked, but one of the easiest to prevent is through unnecessary services.

When you connect to a server, you can see the list of open ports on that machine. An open port is one that has been left accessible, which usually means it needs to be exposed so another computer can use it. But what if you do not need that port? Why leave it open?

Services such as Telnet (Telnet is a protocol that allows you to connect to remote computers called hosts over a TCP/IP network such as the Internet), Secure Shell (SSH) (SSH provides a secure encrypted connection between two hosts over an insecure network), and debug interfaces (debugging is the process of detecting and removing existing and potential errors, also called "bugs," in a software code that can cause it to behave unexpectedly or crash) are often left open by accident because they are not required for regular operation.

A device with open ports with services that are not strictly required for operation may be at risk of an attack. Services such as Telnet, SSH, or a debug interface may play an essential role during development but are rarely necessary for production. By following a few basic principles and doing some simple evaluation, you can significantly reduce the attack surface on your device. Standard (but unnecessary) services include Telnet, SSH, or a debug interface. Ideally, each device in your network should use only those services required for its operation. Using IoT devices with a security model that does not consider their inherent risk, and does not follow an established framework to secure them, can increase the likelihood of exploits for malicious actors.

## 6.3. Keep Your Devices Updated

When it comes to security, there are a lot of things that need to be taken into consideration. The most important one is the update process for your IoT devices. Any credible manufacturer in the security area is typically up to date with patches and updates.

This means these devices can receive software updates and patches from their manufacturers. Vulnerability management is a crucial element of IoT security. It ensures that the integrity and availability of your devices are protected. It also provides transparency into the status of your devices and enables you to respond quickly to issues or attacks.

IoT devices have a lot to remember. They must interact with the Internet and other devices, protect against traffic hijacking and denial-of-service attacks, authenticate users or third parties accessing your services, enforce access control policies, detect and respond to configuration changes, collect and process data in different ways based on usage patterns, and handle system failures and recover gracefully when they cannot be fully recovered.

One of the essential parts of keeping your IoT devices safe is staying on top of patches and updates. You cannot just set your device and forget it—you have to check in now and then to ensure it is still secure and working correctly. Your security depends on it!

IoT devices are vulnerable to threats, including malware and botnets. But when you are up-to-date with patches and updates, you can protect yourself against known vulnerabilities that hackers might try to exploit. And since many IoT devices do not have Internet connectivity, they will not be able to receive software patches or updates on their own—so you need to do it for them!

This sounds like a lot of work, but it does not have to be! Follow these easy steps:

1. Open your browser (or whatever device you want).

2. Go online at least once per day so that any new patch notices will appear in your inbox.

3. Click "yes" when asked whether or not you want to install the update because who wants bad guys messing around with their stuff.

# 6.4. Cutting Out the "Middle Man"

When a device communicates in Plaintext (a plaintext file is a document that contains no formatting, images, colors, or other types of markup), all information being exchanged with a client device or backend service can be obtained by a "Man-in-the-Middle." Plaintext is a collection of symbols, numbers, or letters we can understand. It can be as essential as any piece of text you might create using a word processor or more advanced for coding purposes. The ciphertext is the randomized version of plaintext produced when an encryption algorithm changes things. Decryption algorithms change the ciphertext into something we can understand to match what was entered originally.

Man-in-the-Middle attacks happen when a hacker installs an application or hardware device between you and the website or other service you are trying to communicate with. The attacker can read everything that passes through that device, store it for later use, and change it so other websites will see different information than what happened when you signed in. Anyone capable of obtaining a position on the network path between a device and its endpoint can inspect the network traffic and read the plaintext communication, potentially obtaining sensitive data such as login credentials. A typical problem in this category is using a plaintext version of a protocol (e.g., HTTP) where an encrypted version is available (HTTPS). A secure URL should begin with "HTTPS" rather than "HTTP." The "S" in "HTTPS" stands for secure, which indicates that the site is using a Secure Sockets Layer (SSL) Certificate. This lets you know that all your communication and data are encrypted as it passes from your browser to the website server. Before using a pay portal, it should always have HTTPS leveraging TLS certificates to secure all your data as these are passed from your browser to the website server. Encryption is a process by which information is protected from unauthorized use.

In Man-in-the-Middle attacks, an attacker acts as a man in the middle between two parties and relays messages between them. This can result in intercepting or altering messages without either party realizing they have been compromised. It is necessary to use encryption to prevent data from being intercepted by a Man-in-the-Middle attack. Encryption consists of scrambling the data sent over the network and then unscrambling these at the other end. The most common method of encrypting data is Advanced Encryption Standard (AES), which uses 128-bit blocks. This means that the algorithm generates a key for every 64 bytes of data to encrypt each block with a different key (AES defines a block as 16 bytes). Therefore, one 64-byte block would need to be encrypted again by 64 + 64 = 128 bits. The longer the key, the more complex, and the more challenging to crack. When you use this encryption method, the encryption is usually performed in hardware. The hardware used for this purpose is a cryptographic processor, also used in other security applications such as digital signatures.

Even when data are encrypted, weaknesses may be present if the encryption is incomplete or misconfigured. For example, a device may fail to verify the authenticity of the other party. Even though the connection is encrypted, it can be intercepted by a Man-in-the-Middle attacker. Encryption must be used when sensitive data are stored at rest on a device. Typical weaknesses are a lack of encryption by storing application programming interface (API) tokens. API is a way for different software programs to communicate with each other. Usually, an API integrates two proprietary software packages, allowing them to work well together and share data effortlessly. API tokens allow a user to authenticate with cloud apps and bypass two-step verification or credentials in plaintext on a device. Data transmitted over a network should also be encrypted to prevent eavesdroppers from intercepting it. Encryption is vital for network security and data privacy. However, it is not enough to merely encrypt the data; these must also be delivered securely over the Internet.

The IoT is a growing industry with major security challenges. In addition to the growing number of devices with embedded chips, it also includes cloud-connected systems such as home assistants like Alexa and Google Home. A single device (or even one collection of devices) may be connected to a hacker's computer through an entry point that uses the same connection used by hackers to connect to various websites. In many cases, this is called "hacking," which can compromise data transmitted over the network.

**FIGURE 6.1**  Peter installs a doorbell camera.

Illustrated by Phillip Wandyez.

*After hearing about a break-in to a business across town, Peter decided to beef up his own security. Peter found an IoT doorbell camera system that allowed him to view on his phone who was at the front door of his store. "This is all really nifty and high-tech, I am definitely secure now," he said to himself. Of course, by now we should know that this was not the case! In fact, when the doorbell system was being installed (by the nice gentleman in the hoodie from the café), Peter said, "Yeah, sure whatever, the default settings are fine." As it turns out, this included the username and password, which was later used by a hacker to gain access to Peter's*

*business network.*

# 6.5. **Practice Good IoT Cyber Hygiene**

Most IoT devices are effectively general-purpose computers that can run specific software. This makes it possible for attackers to install their software with functionality that is not part of the normal functioning of the device. For example, an attacker may install software that performs a distributed denial-of-service (DDoS) attack. A DDoS attack is a malicious attempt to disrupt the regular traffic of a targeted server, service, or network by overwhelming the target or its surrounding infrastructure with a flood of Internet traffic. Limiting the device functionality and the software it can run limits the possibility of abusing it. For example, the device can be restricted to connecting only to the vendor's cloud service. This restriction would make it ineffective in a DDoS attack since it can no longer connect to arbitrary target hosts.

A code is typically signed with a cryptographic hash to limit the software a device can run. Since only the vendor has the key to sign the software, the device will only run software distributed by the vendor. This way an attacker can no longer run arbitrary code on a device. A hash is a mathematical function that converts an arbitrary length input into an encrypted output of a fixed length. Hashes are used for many things, including securing database passwords, sending messages securely so that only the intended recipient can read them, keeping track of files on a computer hard drive, and encrypting and decrypting information.

Many devices have minimal security capabilities as part of their hardware or firmware design. For example, they may be isolated from the Internet, contain built-in firewalls, require user permission and authentication to access certain functionality, or include encryption by default. Users should only install software directly from a known source (such as the device manufacturer or another official distributor). The devices connected to the Internet that tend to be most easily abused in cyberattacks are those with exposed or non-protected ports and servers. These devices can be protected from cyber threats by limiting their functionality so they cannot perform potentially malicious acts.

In addition to having a secure password, it is also important that devices require user verification at startup so that only authorized users may access them. The vendor receives input on potential vulnerabilities, develops mitigations, and updates devices in the field. Most security vendors have a process to handle security issues adequately. The consumer mainly perceives the vendor's security posture as improved communication with the vendor about security.

In the case of privacy protection, the vendor plays an important role. Other than an external attacker, the vendor or an affiliated party may be responsible for a privacy breach. Without explicit consent, an IoT device vendor or service provider could gather information on consumer behavior for purposes like marketing research. Several years ago, a neighborhood was targeted due to a heavy volume of unsecured baby monitors. Hackers gained access and strung their capabilities to create a mini server for illegal activities. Many IoT devices store sensitive information like passwords and video and audio recordings. If an attacker accessed this information, it could amount to a severe privacy violation. IoT devices should handle sensitive information correctly, securely, and after the consent of the end-user of the device. This applies to both the storage and distribution of sensitive information. As consumers adopt new IoT devices and services, their expectations around privacy and security will increase. IoT devices offer conveniences such as remote monitoring and control but can also collect sensitive information about your home life. These devices should be designed to protect consumer privacy following established national privacy laws.

Most devices do not have logging or alerting functionality, so any security problems will not be reported. IoT devices are often used by people who are unaware of and powerless to stop security vulnerabilities and attacks. If you think these issues only affect your device, it is essential to realize that devices connected to or within a network are not isolated. One compromised device can expose others, putting all Internet-connected devices at risk. The ability of an attacker to enter into a device and manipulate its functions or disable it renders the owner helpless in defending it.

Device infection is hard to detect, and because most devices have no alerts or logs, owners would have almost no way of knowing about a compromised device. To address this situation, Internet of Things Security Recommendations [INT 1–12] have been developed by the European Telecommunications Standards Institute (ETSI) [1] to put in place better security practices for IoT devices. While these recommendations are designed with a target architecture model ("Narrow Protocol-Level Proximity"), they can be applied as best practices independent of any particular IoT system design.

IoT devices are everywhere, and they are making our lives easier. But they can also be a security risk if you do not take the right precautions. Here are the best practices for keeping your IoT devices secure:

- Stay updated: Update your software regularly to ensure it is always up to date with the latest security patches.

- Use encryption: Encrypting your data will help prevent these from being stolen in case of a breach.

- Automate security with anti-virus software: Anti-virus software can help keep your device safe from malware and other threats.

- Keep tabs on all connected devices: Know what devices are connected to your network and where they are located so you can monitor them closely if they start acting suspiciously (like constantly trying to access new websites).

- Take advantage of authentication tools: Authentication tools like two-factor authentication (2FA) can help add an extra layer of protection against hackers trying to break into your accounts by requiring a second step when logging into accounts or accessing sensitive information like bank accounts or social media profiles.

- Regularly audit actions on your network and disable unused devices: Check for unusual activity.

# 6.6. **Conclusions**

Today, we are facing a new threat as cyber attackers attack the most popular devices in our homes and offices, exploiting security vulnerabilities to obtain access and control over these devices. Attacks can include ransomware attacks that hold a device hostage until a ransom payment is made or malware that creates a botnet of devices to send spam or launch DDoS attacks on websites. The IoT is the network of physical devices, vehicles, buildings, and other items embedded with electronics, software, sensors, and network connectivity to enable these objects to connect and exchange data.

For most of the above security categories, it is difficult for a nontechnical user to evaluate whether a device meets security requirements. However, user interaction can, by definition, be perceived by the end-user, so the consumer can evaluate how well a device performs based on user interaction. By providing a simple, secure configuration for their devices, vendors can encourage the secure deployment of their products. User interaction is one category of security where it is pretty easy for ordinary users to evaluate whether they like or dislike a product. It is therefore easy to use user interaction to evaluate security protections on the networked products they use daily. User interaction with a device is a crucial criterion for the end-user to evaluate whether it meets the requirement for security. The interaction between the user and a device is an essential part of the security of that device. Some devices may have vulnerabilities that could only be effectively exploited by interacting with them incorrectly, such as using a default password instead of changing it.

Additional information is needed to understand how easy it would be for an attacker to exploit security weaknesses. It is vital to ensure that the user can use the device without difficulty. That way, defaults do not lead them to make poor choices, such as using a common password. This is true for any device, especially in the IoT, where many devices are designed for simple interaction via mobile devices. Vendors must consider how users interface with these IoT devices to provide security for this quickly growing area. User interaction is essential to ensure that implemented security measures are activated and correctly used. If you do not know how to use it, it is useless if it is possible to change the default password but the user does not know or cannot discover the functionality. The user is also an essential category in the security of the system, as a user can either be an attacker or a potential victim. The user can accidentally open an unsecured Bluetooth connection or use a default password to the device set by the manufacturer, etc. Security can be complicated and confusing, but it does not need to be.

All devices with IoT technology are exposed to a security threat. The most important aspect of making a device secure, however, is ensuring that users are interacting with the device on an appropriate level. Knowing if the user is behaving "correctly" is a must for any business today. This is especially true with intelligent home appliances, wearables, IoT devices, and more. Securing these devices is imperative to ensure that there are no backdoors or vulnerabilities that could be exploited.

## Reference

1.  ETSI, 2022, accessed October 7, 2022. https://www.etsi.org/.

# 7

# Product Security for Entrepreneurs Selling Digital Products or Services

Kenneth G. Crowther and Dale W. Richards

## 7.1. Introduction

This chapter is for entrepreneurs planning to sell products composed of programmable code or selling a service that affects such digital products. The subject of building cybersecurity into products or services is called "product security."

For entrepreneurs selling digital products or services that support digital products, building cybersecurity into those products may seem either like something needed in the far distant future or an immediate threat that is too hard or too expensive to tackle. Typically, the demand for product security by customers is highly nonlinear—that is, customers do not care about security until they do; at which point, it is too late to build security. Product security is one area in which small adjustments now to demonstrate "due care" yield significant value as customers begin to understand and demand security in their products.

Traditionally, digital entrepreneurs start out by creating what is called a minimum viable product (MVP) that contains only those features that are necessary to create a holistic value offering their customers deem worth buying. This startup stage of the company is characterized by experimentation and feedback collection. Often entrepreneurs are "bootstrapping" or paying for product development out of pocket. Funds are limited. Resources spent need to bring value so there is a lot of pressure and temptation to deprioritize security as something needed later on, perhaps as the company transitions from the startup stage to the growth stage, or perhaps as the product features are validated and entrepreneurs know what their customers need.

However, cyber threats to small businesses are real and product security can be much less expensive when addressed at the outset. It is our goal to outline a basic set of practices entrepreneurs can implement to shift security earlier in their company's growth.

## 7.2. Flaws in Digital Products Can Be Expensive

Overall, the process of fixing a security flaw after a product is released is expensive. Some estimate the "create the patch" part of this process between $400 and $28,000 for Web applications [1], which is a small part of the overall patching process. For products that are delivering value to customers (e.g., operational technologies like logistic management systems, resource sharing apps), a fix that would take a designer 4 h during initial development might require around 1,000 h across a product/integration team to fix a system (to receive, triage, validate, analyze, fix, work with customers for deployment, disclose, etc.). Cobb provides an instructive example in his analysis of the security flaw in the Chrysler Jeep [2]. While the development time to fix the flaw was within the Higgins range [1], the flaw ultimately cost Chrysler in excess of $1 billion due to the large number of Jeeps and the customer interactions necessary to preserve reputation as the Jeep vulnerabilities became public. The cost of fixing security when a product is in operation is orders of magnitude more expensive than during development.

Pretending that product security is not necessary because "we cannot afford it" or "it is something big companies do" or "we are not big enough to draw attention" is not a solution for real threats. Small businesses are prey to cyberattacks. In fact, 75% of data breach victims are small businesses [3]. Sixty percent of small businesses close their doors within six months of a cyberattack [4]. Many modern digital products rely on cloud-connected systems, and the small business is responsible for its maintenance and sustainment—which include security. The average cost of a data breach on a small business is $120,000 [5].

## 7.3. **Shifting Security Earlier**

Software entrepreneurs need to develop a mental model and intuition for how much effort is saved when they attempt to think about security early. Consider the IBM study on the cost of finding and fixing defects in software. According to their summary of several hundred companies (Figure 7.1) [6],

- It costs 6× to fix a bug during development as during design.

- It costs 15× to fix a bug during post-development testing (e.g., final testing) as during design.

- It costs 100× to fix a bug after software deployment compared to during design.

**FIGURE 7.1** The costs of fixing flaws increase as you move closer to production.

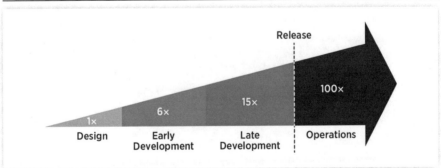

The challenge, of course, is that requirements are highly uncertain during the early design and requirements specification phase.

Therefore, concepts such as Agile and DevOps are emerging. The idea behind Agile is that you try to create a "working product" as early and quickly as possible. The working product provides a chance for the product owner, other stakeholders, and end-users to provide feedback on requirements and priorities earlier in the process. This feedback means that the upgrades to the working product are more likely to reflect real user needs. DevOps is similar, except that it enables individual features to be released to end-users on the order of days instead of being tied to releases resulting from each sprint or other project schedules. In return, product teams can get feedback on individual features from end-users immediately.

**FIGURE 7.2**   Proud Peter.

Illustrated by Phillip Wandyez.

*Peter was in a great mood. He had just closed a big sale of his new high-tech product. But shortly after the sale, the customer called up Peter on the phone. The customer was furious because they discovered a serious security vulnerability in the product. "The cybersecurity of your company is your problem, not mine," said Peter. The*

*customer threatened to disclose the vulner-
ability if Peter did not address the issue.
However, Peter did not really know what to
do with the information anyway as he did not
have a good product security process in
place. The customer decided to disclose the
vulnerability, which ruined the reputation of
Peter's company, causing Peter to miss out
on a substantial amount of revenue.*

## 7.4. **A Basic Security Approach**

What should digital entrepreneurs actually do to incorporate cybersecurity into their products earlier? How can this be achieved without
blowing budgets and killing the business before it is born? There are
some basic activities that digital entrepreneurs can adopt to weave
security into their products from the get-go. For this chapter, we will
outline three pillars of basic product security. These three pillars
should be thought of as continuous and iterative (Figure 7.3).

**FIGURE 7.3** Three pillars of product security (threat modeling, testing, and
sustaining) constitute a place for an entrepreneur to begin their journey toward
building secure digital products.

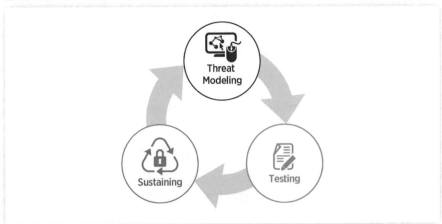

Looking a little closer at the three pillars, here are six activities and when the activities should be performed (Table 7.1):

**TABLE 7.1** Product security activities and when they should be conducted.

| Pillar | Activities | When to do |
|---|---|---|
| Threat modeling | Initial threat modeling | During early product design as user experiences and requirements are defined |
| | Ongoing threat modeling | During each sprint |
| Testing | Automated testing | On each build |
| | Manual and penetration testing | On a regular cadence, as defined by business needs |
| Sustainment | Documenting and training | On a regular cadence, as defined by business needs; with each new team member |
| | Component monitoring and vulnerability management | During each sprint |

Remember this chapter is about product security for building secure digital products and not traditional network or data security. These pillars and activities are derived from existing standards in product security. It can be easy to drown in the complexity of long standards, but we mention some existing standards here for your reference. We will mention these standards throughout the chapter, but we recommend that entrepreneurs familiarize themselves and their teams with at least one of these standards (they all promote similar practices and activities):

- NIST SSDF—The National Institute of Standards and Technology (NIST) Secure Software Development Framework (SSDF), or NIST Special Publication 800-218, defines a set of practices to integrate into software development so that an organization can express their security requirements to third-party suppliers and developers and acquire software that is more secure [7].

- OWASP SAMM—Open Worldwide Application Security Project (OWASP) Software Assurance Maturity Model (SAMM) describes strategies, practices, and activities in a way that an organization can assess their existing posture for developing software that is secure and for which can be communicated a security assurance case [8].

- BSIMM—Building Security In Maturity Model (BSIMM) is a framework that supports a set of services by Synopsys to evaluate and benchmark product security activities for building security into software and digital products. Benchmarks provide a foundation to measure and compare practices across verticals for developing a targeted improvement strategy [9].

- ISA SDLA—The International Society of Automation (ISA) Security Compliance Institute offers certifications of the Secure Development Lifecycle Assurance (SDLA) for companies that develop industrial automation and control systems. These align with the ISA/IEC 62443-4-1 standards for secure development of industrial control or operational technology [10].

A note on maturity: The frameworks listed here are oriented around maturing security practices. However, DevOps practitioners focus on building capabilities—not on maturity [11]. These frameworks contain product security standards that might be too much for the entrepreneur. However, though some of these frameworks describe themselves in terms of maturity, they communicate many useful capabilities that early-stage product teams can implement into their development process.

## 7.5. Threat Modeling

Threat modeling is the process of thinking through your data flows to consider ways that someone might violate the availability, integrity, or confidentiality of data. While you think through these data flows, you will discover weaknesses in your software or implementation design and architecture that might impact your customers. The earlier these potential weaknesses are discovered, the less costly it is to design fixes, controls, or mitigations to ensure that your data are secure as these move throughout your entire product.

Threat modeling should be undertaken during the MVP design. Even early-stage products have a minimally viable set of features.

Those features involve the uptake and processing of data. Getting an experienced cybersecurity professional to model the threats in these features is an inexpensive start to early-stage product security.

Threat modeling fits into both Agile and DevOps frameworks and helps to reduce the uncertainty of security in the end product if they can be deployed early. For example, during the first sprint, you would construct a data flow diagram (DFD) and identify some potential threat susceptibilities associated with the working product. The security controls become user stories, tasks, or requirements for the team. Your velocity then reflects the development of features and security requirements. In general, the first threat model will require about three 1-h meetings, and testing will take several hours to set up and understand. However, during a typical sprint afterward, the threat modeling updates can be time-boxed (e.g., to a half-hour meeting) and the security testing will take minutes to perform. Shifting security early and embedding it with your development process can result in significant cost savings down the line and preserve your reputation as you grow.

In the NIST SSDF model, these are practices in the Produce Well-Secured Software (PW) practice group. The OWASP SAMM model includes activities titled Threat Assessment, Security Requirements, and Security Architecture and provides the foundation for Secure Deployment, Architecture Assessment, and Requirements-Driven Testing. The Synopsis BSIMM model includes practices titled Attack Models, Security Features and Design, Standards and Requirements, and Architecture Analysis and provides the foundation for Security Testing, Penetration Testing, and Software Environment.

As product teams add new features during ongoing sprints, each new feature should be viewed through the lens of cybersecurity. Threat modeling should be applied during each sprint to all new features.

Threat modeling can be done early. Threat modeling is focused on design, architecture, and communications. It is important to understand both the general process of threat modeling but also to understand the philosophy and value of threat modeling. Threat modeling will have the highest impact on security at the lowest cost if your organization embraces the philosophy and values of the Threat Modeling Manifesto [12]. The values of the manifesto align with modern Agile and DevOps practices:

- Engender a culture of finding and fixing design issues over checkbox compliance.

- Enhance people and collaboration over process, methodology, and tools.

- Create a journey of understanding over snapshots of security or privacy.

- Do threat modeling over talking about it.

- Refine iteratively over any big ceremonial single delivery.

Early startups should embrace and integrate these concepts instead of excessive planning. Even biweekly short check-ins on security will enhance the continuous cybersecurity journey and pay dividends on security. It is important to think of this as a continuous and iterative journey and not a big expensive milestone.

The process of threat modeling is:

1. Map out the expected data flows (or the actual data flow if you are further along). This is typically done by getting the team together to develop a DFD. The purpose of the DFD is not to comply with a standard, but to ensure that the entire team has the same understanding of how data move throughout the data stores, data transmissions, data processes, and trust zones that constitute the software.

2. Use general attack patterns to brainstorm threats (ways to ruin the value of the product). For example, STRIDE provides six generalized attack patterns to help you brainstorm: Spoof, Tamper, Repudiation, Information Disclosure, Deny Service, or Elevating Privilege. Alternatives to STRIDE might include brainstorming for violations of security principles, such as Smith [13], Benzel et al. [14], or Harrington [15]. The quality assurance engineer or the product manager is sometimes just as

capable as the security engineer at identifying these threat susceptibilities.

3. Prioritize what can be done in the design and architecture to secure your product against threats. STRIDE is a simple framework because each attack pattern has a method of security. For example, you fix spoofing by strengthening authentication, and you fix tampering by strengthening integrity. This could result in a list of controls, but not all aspects of your software create the same risk. You will want to prioritize those threats that could create the most damage to your customers and fix them in the earliest sprint possible.

4. Iterate when things change or periodically. Just like product development, threat modeling can scale to any timebox that you provide. The point of threat modeling is to develop a continuous attention to security—not to have a singular, solved, and absolute list of security controls.

There are some tradeoffs that we need to make here. For example, using a specific methodology like STRIDE provides a common language and understanding for groups to participate. The methods are only as good as the people observing them. Product security is a team sport and cannot be given to any specific person. In a startup, the quality assurance (QA) engineer might have a very important role in executing threat modeling throughout the product development iterations, but they will not be successful without working collaboratively with the test automation engineers and those who architect and maintain the continuous integration, continuous development (CICD) pipelines.

A classic example of threat modeling is from Amanda Rousseau [16]. She describes the waste that goes into excessive security jargon, false positives of poor security tests, etc. when what is really needed is a better security mindset and some innovation/creativity on the part of the developers and architect. Her example involves getting free pancakes. Consider following the process of threat modeling outlined above:

- Step 1. Map out how you enter the pancake restaurant, how you are greeted by the host, how you are identified and accepted as a valid customer, how you are seated, how you meet your waiter, how to order food, receive food, eat food, and pay for your food. This is the data flow architecture of the pancake restaurant. This is "understanding your system."

- Step 2. Consider methods for getting free pancakes. You might think of scenarios from the simple to the extremely complex. For example, a simple scenario might be to pretend to be an online pick-up customer and just steal pancakes that someone else ordered, or perhaps to pretend to go to the bathroom after you have eaten enough pancakes and just leave. A complex scenario might include a realistic threat of a bioweapon and the antidote only after you eat pancakes. This is "a threat mindset" or "understanding what could go wrong."

- Step 3. Flip perspective from trying to get free pancakes to now you are the restaurant owner. You consider what controls are possible for each scenario and begin to decide what you are willing to do to protect both your revenue and your relationship with customers. These decisions now represent your security thinking and attitude toward risk. There are very few right or wrong answers, security is mostly a journey through tradeoffs. This is "understanding what you will do."

- Step 4. Do not pretend that you thought of everything. Prepare to be wrong. Realize that new fraudulent scenarios will continuously emerge. As a product maker you must revisit and reassess security as your company grows and your customer demand and product requirements shift.

## 7.6. Testing

Testing is the process of verifying the implementation of security controls but also searching for common weaknesses or vulnerabilities that are easy to recognize. There are lots of tools that are available to automate the process of many common tests. The earlier you start testing the more likely you will fix problems in a way to avoid delayed release of the product. Delayed product release can cost market share, which can translate into sustained revenue loss across the lifetime of the product. Numerous automated security testing tools can provide insights as your team develops software. Each automated product has a different niche in which it resides. Increasingly, high-quality open-source software projects are aligning with organizations that donate testing time/equipment so that the users of those open-source libraries benefit from using software components that have been tested and evaluated by the community.

*Software composition analysis (SCA)*—We do not build software or any digital solutions from scratch. SCA helps us to understand our "software bill of materials" (SBOM) so that we quickly identify known vulnerabilities in the libraries or components that we are using. Most vulnerabilities found through SCA can be fixed through an update or patch to the component that contains that vulnerability. Not all vulnerabilities found are exploitable (e.g., we might use only parts of the functionality that are not vulnerable or we might not allow access to the component). Not all vulnerabilities can be fixed (especially in firmware), but we can institute compensating controls and ensure that any known vulnerabilities are not exploitable. In addition to SBOM creation, vulnerability identification, and remediation recommendations, many SCA tools also provide a process for understanding the licenses associated with component reuse. Some software are governed by permissive licenses that enable the reuse, augmentation, and commercialization of software that contain those components. Examples of permissive licenses include Berkeley Software Distribution (BSD), Massachusetts Institute of Technology (MIT), or Apache. These provide substantial permissions to reuse. However, other licenses might require that software built on existing components enable customers to understand composition, exchange core libraries, or even see the source code. Understanding and complying with licenses is not a security issue but is a benefit of SCA. A typical Web-based service potentially uses more than 500 software libraries. SCA provides a quick and cost-effective way to avoid security or licensing issues from reused components.

*Static application security testing (SAST), sometimes called code-based analysis*—There are different types of SAST that run queries against your code and help you trace the source of vulnerabilities and fix them. Good SAST helps you coordinate across teams, prioritize findings, trace findings to the source in the code, fix flaws, and train to recognize those flaws in the future. A SAST tool queries the code in such a way that it mimics human reasoning about the code trying to find potential weaknesses that could be exploitable. There are several open-source and commercial tools. Most open-source tools are quite specific in the scope of what they can check. For example, some open-source tools look at the OWASP Top-10, CWE Top-25, Injection Flaws, or specific cases. One would need to select tools and build a toolbox that is specific to your needs. Many commercial tools integrate across numerous query engines to check for many types of vulnerabilities in addition to typical coding errors, such as hardcoded secrets, exposed keys, and language-specific run-time errors. SAST tools vary widely in terms of accuracy, false-positive rates, user experience, integration with other tools, federate access, and several other features. It is wise to always run a proof-of-concept with samples of software that you understand before signing a license agreement.

*Dynamic application security testing (DAST)* is similar to SAST, except that it runs the queries during the execution of the code, instead of just running the queries on the code itself. The result is that DAST can give you a lower false-positive rate for things like injection vulnerabilities and overflows because it queries the actual workload deployed and executed.

*Fuzz testing or fuzzing* is a process of systematically providing malformed, invalid, unexpected, or random inputs to a software program. While it can be computer-resource intensive, it is typically fully automated and able to find exceptions, crashes, failed code-insertions, etc. This type of testing is becoming increasingly important due to the large input ranges that are designed into modern software. Fuzzing results are typically true positives and come with evidence of precisely what caused the failure, which provides a clear path to resolution.

*Regression testing* is a process of rerunning historical test cases because the software has changed. As a product maker matures, they should continuously improve the automation of regression testing so that the cost of applying security updates or patches is decreased. For example, if you have 500 component dependencies, your average frequency of vulnerability discovery and patch release in your dependent software is about two to three new vulnerabilities/security patches per month. These vulnerabilities can accumulate quickly if you are unable to upgrade your dependencies and deploy new versions of digital products.

Automated security testing should be done as the code is written. It should be a continuous, low-cost activity that is done in small iterations of code development. Depending on the deployment mechanism of the scanning tools, sometimes small pieces of code can be checked or the tool can help point out the differential. The others will sometimes use two licenses of major software products, one license for small segments of code and the other license integrated into the CICD pipeline so that it scans the entire code base after a merge or commit.

*Penetration testing* is a generic term used to describe the process of a security expert trying to discover methods to exploit your digital product. Penetration testing is sometimes called pen testing or ethical hacking. These are not automated and can be expensive. Nevertheless, they are an important part of a testing regime. Typically, pen tests should be scoped clearly (e.g., this pen test focuses on spoofing identities and gaining unauthorized access; the next pen test focuses on exfiltrating stored data or credentials). Typically, at least one pen test is done on a product after 95% of the features are developed for your first release, but before you have released the product for general use. The product owner should schedule pen tests at a regular interval (e.g., tightly scoped pen tests once per quarter or a general-scope pen test annually). Some companies are trying to crowd-source pen testing by creating a separate product with fictitious data and some exposures, and then offering a bug bounty for those who provide evidence of exploitations that are not yet known.

Testing is increasingly being required by governments and industry bodies. In the US, a recent executive order (EO 14028 *Improving the Nation's Cybersecurity*) directed the NIST to create standards for software testing. NIST published IR 8397 *Guidelines on Minimum Standards for Developer Verification of Software* [17], which provides high-level summaries of several testing methodologies that are recommended to everyone and will be required for critical software.

## 7.7. Sustainment

Good testing throughout the lifecycle establishes a pattern that can be driven toward security sustainment. Automated testing as the code is written is the foundation of sustainment. At the core of sustainment is first a realization that continuous product development will be required throughout the lifetime of the product. There is no such thing as perfect software, and in an age of security exploitation, one will need to sustain and support software until a published "last day of support." This is a new mindset for product makers because traditionally, we embraced the idea that once a product is commercialized, we are done with support except for warranty requests. In our modern age of digital products with dependencies on existing software libraries, this model has shifted, and continuous sustainment is expected. This section does not describe how to make this cultural and managerial shift to a new sustainment model but describes some common practices to improve our capabilities to identify and remediate security issues as they arise.

Sustaining is necessary because digital products are never finished. Flaws or vulnerabilities are likely to exist and be discovered. Designing a product in a way that makes updating and fixing flaws easy and cost-efficient will result in an ability to iterate continuously through product cybersecurity so that your security will grow at a rate consistent with the growth of your business and emerging attack surface.

*Documentation and training*—Secure deployment guidance is documentation that communicates to a customer or internal stakeholders the security context and decisions that were made during the development, integration, testing, and deployment processes and provides recommendations on how to configure, integrate, and maintain defense in depth for typical product deployments. This is an important opportunity to communicate that cybersecurity is a shared responsibility and helps to demonstrate what risks remain (or are transferred to the customer by a third party). For example, most Amazon Web Services (AWS) publish extensive guidance on how to configure, build, integrate, and secure their products. This guidance is consistent with their published "shared responsibility model" [18]. Moreover, they provide automated and semi-automated audit and security checks that help customers to implement guidance. Most products have installation, operations, and user guides and manuals. The trend is to add a cybersecurity chapter that communicates this security context and recommended guidance at the appropriate level of generalization. Most secure deployment guidance includes a high-level description of defense-in-depth controls built-in, product hardening guidelines, secure operations guidelines, secure disposal guidelines, and clear instructions on how to control and manage access accounts. This documentation becomes the foundation of training new product team members and potentially other stakeholders. It can also be used to train current team members on a regular basis as deemed necessary by the business.

*Product security incident response team (PSIRT)*—PSIRT is "an entity within an organization which, at its core, focuses on the identification, assessment and disposition of the risks associated with security vulnerabilities within the products, including offerings, solutions, components and/or services which an organization produces and/or sells" [19]. When instituted properly, PSIRT is tightly integrated into an enterprise's security engineering initiatives and the product secure development framework. In a startup, the functions of the PSIRT need to be carried out by the product development team at a smaller scale. PSIRT is a framework of activities for paying attention to stakeholders across the product-use community and being able to understand and act on what the community finds in terms of security vulnerabilities, product misuse, security flaws, or security incidents that might affect the company's or the product reputation. FIRST.ORG describes six service areas that constitute PSIRT capabilities. These include:

1. Stakeholder management—the identification and communication with stakeholders (e.g., customers, vendors, resellers, security researchers, product teams) and establishing an ecosystem in which finding and disclosing security features can be cost-effective.

2. Vulnerability discovery—these are a set of activities for monitoring various stakeholder groups to discover or to receive discoveries of vulnerabilities found by stakeholders.

3. Vulnerability triage—these are a set of activities to qualify and validate which security discoveries are valid and how important they are to the business or the product reputation.

4. Vulnerability remediation—these are activities to engage the product teams to get security flaws onto development roadmaps and track the process of fixing or creating patches.

5. Vulnerability disclosure—these are activities to take information regarding the discovery, validation, and remediation and responsibly disclose them to stakeholders—while sometimes the focus here is public disclosure (e.g., releasing a Common Vulnerabilities and Exposures (CVE) notice to the NIST National Vulnerability Database) many disclosures are internal or directly to customers, depending on the product and the risk associated with the disclosure.

6. Training and education—these are tasks associated with ensuring that stakeholders understand and are able to execute communication when there are PSIRT issues.

It is critical for early-stage product teams to become familiar with these product vulnerability management practices. Otherwise, they will be left scrambling when a breach occurs. Even the earliest of teams will benefit from knowing what to do in an emergency.

There is a trend to improve the scalability of PSIRT and other incident responses through the use of a standard "incident command system" that has been in development for decades in the larger emergency management community. The ISA Global Cybersecurity Alliance, for example, has created an initiative to get vendors and users of industrial automation and control systems and devices to all conform to the standardized incident command structure [20]. This enables scaling when there are large systemic vulnerabilities, such as SolarWinds or Log4j.

*Third-party component monitoring and vulnerability management*—While the term vulnerability management is used extensively in the network security and operations security space, it is equally as important in the product sustainment space, but has a different meaning. Vulnerability management in the product security space is about monitoring the libraries, frameworks, and components on which your digital product is built and ensuring that you can quickly receive notifications and apply patches from the vendors of the components that you use. This sometimes motivates the need for contracts with suppliers that ensure that they will provide notice when there is a vulnerability or that they will maintain up-to-date SBOMs for use of automated scanners. Typically, most of the software is coded from imported libraries and reused frameworks, and as such, most of the vulnerability in products comes from those components.

Traditionally, much of security was done before release. However, the trend is to shift left and then right. We shift left by paying more attention to the architecture and design of data flows. This is where we have concepts, such as threat modeling that drive zero-trust architectures through the careful design of data stores, data processes, and data flows such that no individual entity is able to violate security and exploit users. We then shift right through enhanced sustainment and continuous monitoring of products. This is where concepts such as threat hunting, in which we take emerging tactics, techniques, and procedures of adversaries, are documented and cataloged according to the MITRE

ATT&CK model [21]. We develop a list of evidence that we would need to see if those techniques are attacking our product, and then we hunt for that evidence. Testing is still a pillar of good security, but with modern DevOps, sustainment is becoming increasingly important.

# 7.8. Conclusions

This chapter has presented the foundations for an entrepreneur to start their product security journey. While there are conceptual hurdles to adopting new practices, the marginal costs of product security are small when the organization embeds security into its culture and managerial processes across the product lifecycle, that is, when thinking of requirements, they consider security requirements; when planning an architecture, they do some threat modeling; when planning versioning and change control, they integrate with automated security testing; and most importantly when budgeting for product lifetime, they remember to budget for security sustainment. The key to remaining lean while doing product security is to take small momentums for every segment of code and every feature to think about security.

Product security is ultimately a shared responsibility between the customer and the digital product maker. No software engineer can solve every security flaw, and no customer can blame all operational errors on the vendor. As such, the important part of product security is to begin the journey and try to integrate a secure development framework across the organization and the software development process wherever possible. This means ensuring that you are fundraising for sustainment.

While these pillars (threat modeling, security testing, and sustainment) constitute a strong foundation and first step in a product security journey, they will not solve all security challenges. As your organization grows and your digital products become more complex, you will need to continuously add product security capabilities. Product security is a continuous journey.

# References

1. Higgins, K.J., "The Cost of Fixing an Application Vulnerability," Dark Reading, 2009, accessed December 23, 2022, https://www.darkreading.com/risk/the-cost-of-fixing-an-application-vulnerability/d/d-id/1131049.

2. Cobb, S., "Cybersecurity and Manufacturers: What the Costly Chrysler Jeep Hack Reveals," WeLiveSecurity, accessed December 23, 2022, https://www.welivesecurity.com/2015/07/29/cybersecurity-manufacturing-chrysler-jeep-hack/.

3. Verizon, "Verizon Data Breach Report 2017," 2017, accessed December 23, 2022, https://www.verizon.com/business/resources/reports/2017_dbir.pdf.

4. Galvin, J., " 60 Percent of Small Businesses Fold Within 6 Months of a Cyber Attack. Here's How to Protect Yourself," *Inc Magazine*, 2018, https://www.inc.com/rebecca-deczynski/power-of-purpose-list-impact-driven-companies.html.

5. Grustniy, L., "What Is the Cost of a Data Breach," Kaspersky, 2018, accessed December 23, 2022, https://usa.kaspersky.com/blog/economics-report-2018/15445/.

6. Dawson, M., Burrell, D., Rahim, E., and Brewster, S., "Integrating Software Assurance into the Software Development Life Cycle (SDLC)," *Journal of Information Systems Technology and Planning* 3, no. 6 (2010): 49-53.

7. NIST, "Secure Software Development Framework (SSDF) Version 1.1: Recommendations for Mitigation the Risk of Software Vulnerabilities," NIST SP 800-218, 2022, accessed December 23, 2022, https://nvlpubs.nist.gov/nistpubs/SpecialPublications/NIST.SP.800-218.pdf.

8. OWASP, "Software Assurance Maturity Model (SAMM) Version 2," 2022, accessed December 23, 2022, https://owaspsamm.org/.

9. Synopsys, "Building Security In Maturity Model (BSIMM) Version 13: Foundations Report 2022," 2022, accessed December 23, 2022, https://www.synopsys.com/software-integrity/engage/bsimm-web/bsimm13-foundations.

10. ISA Secure, "Secure Development Lifecycle Assurance (SDLA) Certification Version 3.0.0," 2020, accessed December 23, 2022, https://isasecure.org/certification/iec-62443-sdla-certification.

11. Forsgren, N., Humble, J., and Kim, G. "Accelerate: Building and Scaling High Performing Technology Organizations," IT Revolution, 2018.

12. Braiterman, Z. et al., "Threat Modeling Manifesto," n.d., accessed December 23, 2022, https://www.threatmodelingmanifesto.org/.

13. Smith, R.E., "A Contemporary Look at Saltzer and Schroeder's 1975 Design Principles," *IEEE Security & Privacy* 10, no. 6 (2012): 20-25.

14. Benzel, T.V., Irvine, C.E., Levin, T.E., Bhaskara, G. et al., "Design Principles for Security," Naval Postgraduate School, 2005, accessed December 23, 2022, https://apps.dtic.mil/sti/pdfs/ADA437854.pdf.

15. Harrington, T., "Secure Design Principles: How to Build Systems That Are Resilient Against Attack," Linked-In, 2017, accessed December 23, 2022, https://www.linkedin.com/pulse/secure-design-principles-how-build-systems-resilient-ted-harrington/.

16. Rousseau, A., "Blue to Red: Traversing the Spectrum," Black Hat Europe 2019, accessed December 23, 2022, https://youtu.be/WhSrLk6vWgQ?t=2835.

17. Black, P.E., Guttman, B., and Okun, V., "Guidelines on Minimum Standards for Developer Verification of Software," NIST Internal Report (IR) 8397, 2021, accessed December 23, 2022, https://nvlpubs.nist.gov/nistpubs/ir/2021/NIST.IR.8397.pdf.

18. AWS, "Shared Responsibility Model," n.d., accessed December 23, 2022, https://aws.amazon.com/compliance/shared-responsibility-model/.

19. FIRST, "Product Security Incident Response Team (PSIRT) Services Framework Version 1.1," 2020, accessed December 23, 2022, https://www.first.org/standards/frameworks/psirts/psirt_services_framework_v1.1.

20. ISA, "ISA Global Cybersecurity Alliance Incident Command Systems for Industrial Controls Systems (ICS4ICS)," n.d., accessed December 23, 2022, https://gca.isa.org/ics4ics.

21. Pennington, A. (ed.), "Getting Started with ATT&CK," 2019, accessed December 23, 2022, https://www.mitre.org/sites/default/files/2021-11/getting-started-with-attack-october-2019.pdf.

# 8

# Strategic Startup in the Modern Age: Cybersecurity for Entrepreneurial Leaders

Samantha Bryant Steidle

## 8.1. Introduction

Starting a business in the modern age requires special considerations, compared to past decades, as the startup process has evolved dramatically. For years, the first step of starting a business was writing a traditional business plan. Modern scholars now advocate for a new formula for entrepreneurial startup success. This new formula often combines modern entrepreneurial strategies, tools, and networking. Modern entrepreneurial strategies, such as effectuation [1], design thinking [2], systems thinking [3], and entrepreneurial thinking [3], are complemented by entrepreneurial tools, including customer development [4], business model canvas [5], lean startup [6], and networking through entrepreneurial ecosystems and the builders that help them thrive. In this chapter, we will introduce these modern startup strategies, tools, and networking suggestions, which are truly the foundation for entrepreneurial success in the modern digital age.

What does all this have to do with cyber risks for entrepreneurs? Let us pull it all together in the context of the modern age. The modern age means post-pandemic technological advancements with interconnected and complex challenges. Increased levels of automation, artificial intelligence, and machine learning result in increased complexity in terms of cyber risks. For a startup to compete with larger corporations, it must think smarter to find multidimensional solutions through collaborative ecosystem networks. By meaningfully engaging with an ecosystem that includes cyber professionals, cybersecurity risks can be lessened. When entrepreneurs take an ecosystem approach, they are able to better leverage expertise throughout a region.

**FIGURE 8.1**    Cyberattack strikes again.

Illustrated by Phillip Wandyez.

*After a number of failed business ventures, Peter was struck by another wave of inspiration to start a new business. He knew at this point that he should give some thought to cybersecurity, but with everything else he needed to do – talking with customers, talking with suppliers, filling out paperwork, etc. – he just did not have any time (much less, money) to worry about cyber. "I'm not even sure where to start anyway, and besides, what are the chances that a cyber-security incident could happen to me again?" he thought to himself. "I'm not THAT unlucky." But unfortunately, this is Peter the Salesman after all, who is in fact, that unlucky. By failing to plan accordingly as he started and grew his new business, Peter was struck by a cyberattack that threw his whole business venture into disarray.*

## 8.2. **Modern Entrepreneurial Strategies**

In recent years, research on entrepreneurship has expanded dramatically. Gone are the days of combining small business management with the startup process. Today, scholars widely embrace the notion that starting a business is quite different from managing a business. Therefore, the strategies required for a successful launch are also different. While the strategies may be viewed as theory, the magic for entrepreneurs happens when the theory is put into action. For this reason, readers of this chapter would be wise to view the information provided with a bias toward action. This section will serve as an introduction to several modern entrepreneurial strategies, including effectuation, design thinking, systems thinking, and entrepreneurial thinking. These strategies are quite crucial for startup success in the 21st century.

## 8.2.1. Effectuation

Effectuation is a way of thinking, which encourages one to make decisions based on past experiences, accumulated knowledge, and currently available resources [7]. Rather than starting with a predetermined goal using linear and causal logic, effectuation relies on and evolves with innate logic. For example, the entrepreneurial journey inevitably involves exploration, which refines the objectives and often adjusts the path. Entrepreneurs today grapple with new issues, such as cybersecurity threats. According to Sarasvathy, effectual logic is more appropriate for these dynamic threats, which modern entrepreneurs must navigate [7].

In general, effectuation encourages one to start with who you are, what you know, and who you know. The four principles of effectuation are (a) Bird in hand, which encourages value creation based on the resources one currently has access to; (b) Lemonade principle, which emphasizes that mistakes are inevitable but can lead to new opportunities; (c) Crazy quilt, which views new partnerships as opportunities to gain new perspectives and funding because meeting new people often expands who and what you know; (d) Affordable loss, which encourages the individual to only invest the amount they are willing to lose [7]. In general, the individual is encouraged to "begin with a simple problem for which he/she sees an implementable solution—or even something that you simply believe would be fun to attempt" [1]. While using effectuation, "action trumps analysis" [1].

## 8.2.2. Design Thinking

Design thinking is defined as "a process of actions and decisions aimed at producing products, services, environments, and systems that address a problem and improve people's lives" [8]. The central tenets of design thinking are multidisciplinary, human centered, prototype driven, and ideation based. According to Brown and Katz, design concepts are employed as agents of change [9]. The empathy-driven process involves working directly with end-users to understand their pain points and stressors for the purposes of designing a human-centered solution or intervention to address the pain points described. During the process, entrepreneurs ask questions like, "How might we better support rural business owners after COVID-19?" The rigorous methodology also acts as a "mechanism for nurturing future leaders" and "brings creative techniques to the public for the greater good" [3].

## 8.2.3. Systems Thinking

Systems thinking is defined as "a process of understanding interactions and influences between various components in a system to solve complex problems, by addressing every issue as a component of a larger system, rather than an independent aspect with non-related consequences" [3]. The concept is characterized by several key concepts, including:

(a) Viewing and addressing problems holistically.

(b) A mindset of consistent learning, adaptation, and resilience, rather than planning, execution, and rigidity.

(c) Reliance on the synthesis of information and intuition.

(d) The willingness to take accountability for conditions and act to improve them.

(e) An understanding that "meaningful, lasting change requires addressing deep, structural problems over a sustained period."

(f) A small number of high-leverage interventions have a more significant impact than single, isolated interventions [10].

Massachusetts Institute of Technology professor and systems scientist, Peter Senge, published *The Fifth Discipline: The Art and Practice of the Learning Organization* in 1990. In the book, Senge explained that humans tend to focus on what is happening around them simply because it is most observable, failing to recognize the underlying mental models that we all construct to understand the world around us, which influence what is happening on the surface. To illustrate this point, Senge introduced the Iceberg Model of Systems Thinking [11, 12]. The model encourages one to think critically about the reasons for the event or activity. What has changed? For example, if job creation numbers are declining within a region, what has happened that may have caused the decrease? Perhaps the local college discontinued community classes aimed at business startups. Next, the model encourages an inquiry into why this happened. Maybe state budget cuts have forced college administrators to make cuts based on which courses are not financially sustainable. The model now prompts questions about underlying assumptions and beliefs which drive the behavior. Perhaps the college assumed additional funding was not available to support entrepreneurial job creation. The root cause can now more effectively be addressed but would not have been identified if we had not looked beneath the surface. The same process can be used to take a systems-thinking perspective for cyberattack ramifications. However, systems thinking can be used in numerous other ways for entrepreneurs.

Systems thinkers naturally consider how seemingly unrelated issues are interconnected [13]. As Harvard Business professor Vikram Mansharamani has explained, "Breadth of perspective and the ability to connect the proverbial dots (the domain of generalists) is likely to be as important as the depth of experience and the ability to generate dots (the domain of specialists)" [13]. Similarly, one of Google's top recruiters emphasized that the organization values problem-solvers who possess "general cognitive ability" over knowledge related to a specific role [13]. Entrepreneurial systems thinking is critical for addressing complex global challenges [10], such as cybersecurity threats.

## 8.2.4. Entrepreneurial Thinking

Entrepreneurial thinking is "a mindset that emphasizes recognizing opportunity and learning to capitalize on it in a manner unique to the situation" [3]. The mindset involves applying effectual reasoning, or discovery-driven planning, which influences the goals to shift as new information is gained, rather than starting with concrete goals. According to Patel and Mehta, the central tenets of entrepreneurial thinking are collaboration, value creation, discovery driven, and resilience [3]. Modern research has increasingly focused on the higher-order cognitive strategies leveraged by entrepreneurs [14].

How does it all fit together? Interestingly, after Patel and Mehta examined the individual tenants of systems, design, and entrepreneurial thinking, the intersections between the three were analyzed [3]. According to the analysis, entrepreneurial thinking is a mindset that identifies opportunities to create value and resilience through collaboration and human interaction. Once the idea has been identified, the entrepreneurs can use design thinking to explore and refine the problem statement with a multidisciplinary and multistakeholder lens, and ideate for potential solutions or interventions ideally with the end-user while building and testing prototypes of the solution. Finally, systems thinking views the proposed solution through a lens of holistic interdependence, which means that "the parts only have meaning in relation to the entire system" [3]. Informed system thinkers often hesitate to implement interventions before thoroughly understanding the whole system to avoid unintended consequences of a proposed intervention.

According to Patel and Mehta, "when an entrepreneurial thinker attempts to create value through innovation, he or she leverages design thinking to identify new opportunities" [3]. Additionally, "design thinking facilitates the creation of intrinsic value in products or ideas, whereas entrepreneurial thinking is a means of bringing that value to realization" [3]. Finally, systems thinking "harmonizes improvement across an entire ecosystem" [3]. The processes, tools, methods, and theories are often used together as a toolbox for complex problem-solving. Often, the question is which tool or combination of tools is best suited to address the problem at hand.

## 8.3. Modern Entrepreneurial Tools

### 8.3.1. Business Model Canvas

The Business Model Canvas is a one-page visual tool used to describe how an organization or individual "creates, delivers, and captures value" [5]. The nine building blocks of the canvas include the key partners, key activities, key resources, cost structure, value proposition, customer relationships, channels of distribution, customer segments, and revenue streams. It is important to note that when the entrepreneur first fills out the Business Model Canvas, the content is at best an educated guess. For this reason, the processes of lean startup and customer development become a crucial part of validating the business model to increase the likelihood of success.

Cybersecurity could fit as a component in several of the blocks. For example, cybersecurity could be core to the value proposition, customer segmentation, or partnership strategy. Cybersecurity could be a key activity or part of the cost structure. The point is the Business Model Canvas provides various perspectives for value creation both for the customer and the entrepreneur, which increases financial sustainability.

### 8.3.2. Lean Startup

Lean Startup is defined as "an organization formed to search for a repeatable and scalable business model" [15]. In the Harvard Business Review article titled *Why the Lean Startup Changes Everything*, Stanford professor, Steve Blank, described Lean Startup as "a smarter, faster, methodology for launching companies" expected to "make business plans obsolete." He also believes, "if widely adopted, it would reduce the incidence of start-up failure." Similarly, Carl J. Schramm, Former President of the Ewing Marion Kauffman Foundation, advocates for burning the business plan. Customer development is one of the core components of the lean startup methodology.

### 8.3.3. Customer Development

It is a value creation tool that encourages the entrepreneur to consider, "What is the smallest or least complicated problem that the customer will pay us to solve?" [16]. Blank and Dorf emphasized, "there are no facts inside your building, so get outside…. and into conversations with your customers" [16]. While employing customer development, the action involves conducting experiments to test the original hypothesis, which often evolves, based on patterns of new information gained through customer feedback loops.

Previously, we discussed how cybersecurity could be leveraged across several of the boxes in the Business Model Canvas. The customer development process is an ideal method for testing these assumptions. By engaging with enough potential customers and users, the assumptions are further validated. This means the entrepreneur is better able to build and invest in what matters most to the customer.

## 8.4. Modern Entrepreneurial Networking

### 8.4.1. Entrepreneurial Ecosystems

For years, entrepreneurs were viewed as independent agents acting as fringe players, but in recent years, that trend has changed. Today, it is more common than ever for entrepreneurs to take a more collaborative approach to the startup process through what is called entrepreneurial ecosystems. Entrepreneurial ecosystems are defined as "the geographically-bound systems of individuals, organizations, physical resources, social structures, and cultural values that generate new venture activity" [17]. There are numerous benefits for the entrepreneur of taking this more collaborative approach. For example, these communities often provide much-needed support through mentorship, education, funding, policy change, and community. With the digital age changing so rapidly, it is critical for the entrepreneur to both be aware of the ecosystem and plug into this database of resources. Most communities have an established ecosystem that can be initially accessed through either a simple Google search of the "entrepreneurial ecosystem [your city]," by visiting the local Small Business Development Center and/ or plugging into the local technology council. Additionally, entrepreneurs can seek out an ecosystem builder in the local community.

Entrepreneurial ecosystems are widely recognized in the world of startup advocacy, education, and economic development. The concept brings value to entrepreneurs from a cyber perspective, as well. Figure 8.2 provides a visual representation of entrepreneurial ecosystems as a framework for cyber support in a startup context [18].

**FIGURE 8.2**    Entrepreneurial ecosystem for cyber support.

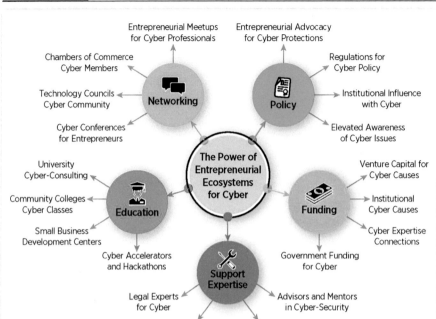

## 8.4.2. Ecosystem Builders

Ecosystem builders are individuals who drive long-term and system-wide change by supporting innovation and entrepreneurship in their region or community through (a) leading recognized startup ecosystem-building initiatives; (b) running entrepreneurial centers and coworking spaces; (c) managing accelerators, incubators, or startup school programs; (d) serving in professional economic development or government roles; or (e) investors and serial entrepreneurs investing in building their local ecosystem [19]. The Kauffman Foundation considers ecosystem building a "new emerging model for economic development in the connected age" through ecosystem mapping [20]. These connected individuals often take part in ecosystem mapping initiatives, which inform them of the major players and resources that support entrepreneurs in each community. Ecosystem mapping is a process that involves developing categories of who is involved in the ecosystem and what role that individual plays. Entrepreneurial ecosystems tend to connect around online and in-person events, which entrepreneurs can and should attend to start networking in the modern age. Search the Internet for an entrepreneurial ecosystem event calendar in your community.

## 8.5. Value of Entrepreneurial Strategies, Tools, and Networking in the Digital Age

Startups operate in a highly resource-constrained environment, which makes accessing this expertise a challenge. The ecosystem approach provides critical support otherwise inaccessible to entrepreneurs. For example, cyber accelerators and hackathons encourage cyber professionals to compete in business challenges. Startup support networks, such as Small Business Development Centers and Technology Councils, provide education and networking opportunities for entrepreneurs. While cyber topics are covered, the far greater value lies in the connections to startup-friendly professionals. They have experience navigating cyber challenges in their own companies or they know the best experts in the field. Institutions of higher education provide human capital in the form of cybersecurity graduates within entrepreneurial ecosystems. Additionally, cyber challenges are often a policy issue due to the regulatory nature of the industry. By engaging regularly with the ecosystem, entrepreneurs also have increased access to policymakers and advocates for the policy change needed. Once an ecosystem cyber network is established, entrepreneurial strategies, such as design thinking and business model canvassing, are much more valuable.

While it may not be immediately obvious, these modern entrepreneurial techniques are ground zero for effectively navigating cybersecurity. After all, entrepreneurs often experience resource-constrained environments. The suggestions offered in this chapter enable a startup to leverage resources, knowledge, and skills throughout an ecosystem, enabling them to achieve startup success in the modern age (Table 8.1).

**TABLE 8.1**  Additional resources for modern entrepreneurs.

| Topic | Resource | Link |
|---|---|---|
| Modern Entrepreneurial Strategies | Kauffman Foundation— watch this video | https://www.youtube.com/ watch?v=8NBnoVrLFPU |
| Modern Entrepreneurial Strategies | Entrepreneurial Learning Initiative—watch this keynote | https://www.youtube.com/ watch?v=r6RG9MfGzLc |
| Modern Entrepreneurial Tools | Strategizer—watch this free video series | https://www.youtube.com/ watch?v=QoAOzMTLP5s&li st=PLBh9h0LWoawqBJk47 ls8XWqaPg8h3WK4S |
| Modern Entrepreneurial Tools | Steve Blank—review this list of startup tools | https://steveblank.com/ tools-and-blogs-for-entrepreneurs/ |
| Modern Entrepreneurial Tools | Strategizer—leverage these free canvas tools | https://www.strategyzer. com/canvas |
| Modern Entrepreneurial Tools | Read *Why the Lean Startup Changes Everything* by Steve Blank (HBR) | https://hbr.org/2013/05/ why-the-lean-start-up-changes-everything |
| Modern Entrepreneurial Tools | Read *The Lean Startup: How Today's Entrepreneurs Use Continuous Innovation to Create Radically Successful Businesses* | https://www.amazon.com/ Lean-Startup-Entrepreneurs-Continuous-Innovation/ dp/0307887898/ |
| Modern Entrepreneurial Networking | Meet Brad Feld—watch this video about startup ecosystems | https://www.youtube.com/ watch?v=zXD5vt0xhyI |
| Modern Entrepreneurial Networking | Small Business Development Center—view cybersecurity resources | https://www.youtube.com/ watch?v=eP4mBftFgXE |
| Modern Entrepreneurial Networking | Read *Startup Communities* by Brad Feld | https://www.amazon.com/ Startup-Communities-Building-Entrepreneurial-Ecosystem/dp/1118441540 |
| Modern Entrepreneurial Networking | Read *The Startup Community Way* by Brad Feld and Ian Hathaway | https://www.amazon.com/ Startup-Community-Way-Entrepreneurial-Ecosystem/dp/1119613604 |

# References

1. Read, S., Sarasvathy, S., Dew, N., Wiltbank, R. et al., *Effectual Entrepreneurship* (Abingdon, UK: Routledge, 2011), doi:https://doi.org/10.4324/9780203836903

2. Johansson-Sköldberg, U., Woodilla, J., and Çetinkaya, M., "Design Thinking: Past, Present and Possible Futures," *Creativity and Innovation Management* 22 (2013): 121–146, doi:https://doi.org/10.1111/caim.12023.

3. Patel, S. and Mehta, K., "Systems, Design, and Entrepreneurial Thinking: Comparative Frameworks," *Systemic Practice and Action Research* 30, no. 5 (2017): 515–533, doi:https://doi.org/10.1007/s11213-016-9404-5.

4. Blank, S., *The Four Steps to the Epiphany*, 5th ed. (Hoboken, NJ: John Wiley & Sons, Inc., 2020)

5. Osterwalder, A. and Pigneur, Y., *Business Model Generation: A Handbook for Visionaries, Game Changers, and Challengers* (Hoboken, NJ: John Wiley & Sons, 2010)

6. Ries, E., *Lean Startup* (New York: Random House Digital, 2010)

7. Sarasvathy, S.D., "Causation and Effectuation: Toward a Theoretical Shift From Economic Inevitability to Entrepreneurial Contingency," *Academy of Management Review* 26 (2001): 243–263, doi:https://doi.org/10.2307/259121.

8. Boni, A.A., Weingart, L.R., and Evenson, S., "Innovation in an Academic Setting: Designing and Leading a Business through Market-Focused, Interdisciplinary Teams," *The Academy of Management Learning and Education* 8 (2009): 407-417.

9. Brown, T. and Katz, B., "Change by Design," *The Journal of Product Innovation Management* 28, no. 3 (2011): 381–383.

10. Feld, B. and Hathaway, I., *The Startup Way: Evolving an Entrepreneurial Ecosystem* (Hoboken, NJ: John Wiley & Sons, 2020)

11. Meadows, D.H., *Thinking in Systems: A Primer* (White River Junction, VT: Chelsea Green Publishing, 2008)

12. Stroh, D.P., *Systems Thinking for Social Change: A Practical Guide to Solving Complex Problems, Avoiding Unintended Consequences, and Achieving Lasting Results* (White River Junction, VT: Chelsea Green, 2015)

13. Mansharamani, V. "Harvard Lecturer: 'No Specific Skill Will Get You Ahead in the Future'— But This 'Way of Thinking' Will," CNBC, June 15, 2020, accessed October 17, 2022, https://www.cnbc.com/2020/06/15/harvard-yale-researcher-future-success-is-not-a-specific-skill-its-a-type-of-thinking.html.

14. Haynie, J.M., Shepherd, D., Mosakowski, E., and Earley, P.C., "A Situated Metacognitive Model of the Entrepreneurial Mindset," *Journal of Business Venturing* 25, no. 2 (2010): 217–229.

15. Blank, S., "Why the Lean Start-Up Changes Everything," *Harvard Business Review* 91 (2013): 63-72.

16. Blank, S. and Dorf, B., *The Startup Owner's Manual: The Step-By-Step Guide for Building a Great Company* (Pescadero, CA: K&S Ranch, 2012)

17. Roundy, P., "Hybrid Organizations and the Logics of Entrepreneurial Ecosystems," *International Entrepreneurship and Management Journal* 13, no. 4 (2017): 1221–1237.

18. Isenberg, D.J., "The Entrepreneurship Ecosystem Strategy as a New Paradigm for Economic Policy: Principles for Cultivating Entrepreneurship," [Paper Presentation], The Institute of International European Affairs, Dublin, Ireland, 2011, accessed October 17, 2022, http://www.innovationamerica.us/images/stories/2011/The-entrepreneurship-ecosystem-strategy-for-economic-growth-policy-20110620183915.pdf.

19. Horn, A. "EshipSeries: So You Think You're an Ecosystem Builder… A Job Description and Training Curriculum for Ecosystem Builders of the Future," 2017, accessed October 17, 2022, https://bthechange.com/so-you-think-youre-an-ecosystem-builder-63c8f2ab9fde.

20. Kauffman Foundation, "Entrepreneurial Ecosystem Building 3.0," 2021, accessed October 17, 2022, https://www.kauffman.org/ecosystem-playbook-draft-3/.

# 9

# Cyber Law for Entrepreneurs

Jennifer Dukarski

## 9.1. Introduction

In the United States (US) and many countries, cyber law is intertwined with the concept of data privacy and data protection. "Data Privacy" refers to the relationship between the collection, storage, use, dissemination, and security of information identifiable or defined as private, the varying public expectations (or not) of privacy, and the attendant legal and political tensions. "Data Security" or "cybersecurity" is the protection of data from any unauthorized access in violation of policy, law, regulation, or rule.

While international privacy law is robust and the US is seeking to become more integrated into the protection of data privacy, cybersecurity often takes a back seat and is approached in a very sectoral manner as the US focuses highly on certain industry segments. This chapter will primarily address the laws applicable in the US.

# 9.2. Federal Laws, Executive Orders, and Regulations

There are not many comprehensive federal laws that directly impact cybersecurity across all industries. Despite this, there are many laws requiring certain sectors to provide for adequate levels of data protection.

## 9.2.1. Federal Laws and Regulations

### Federal Trade Commission Act: Targeting Deceptive Practices

Likely the broadest of federal laws, the Federal Trade Commission Act (FTC Act) is enforced by the Federal Trade Commission (FTC) which uses its enforcement powers under Section 5 of the FTC Act to address unfair or deceptive trade practices.[1] In the terms of cybersecurity, that typically occurs when a company misrepresents its security practices or fails to take reasonable measures to protect individuals' data. When the FTC gets involved, it can either follow an administrative approach in issuing a cease-and-desist order and a consent decree or it may file a complaint in court seeking an injunction and redress for consumers. When reviewing data security issues, the FTC applies a reasonableness standard, meaning that a business must have used reasonable technology and business practices in its operations.

For over 20 years, the FTC has led the enforcement efforts in many areas, including:

- *The COPPA Rule (Children's Online Privacy Protection Act).*[2] COPPA creates a reasonableness standard with respect to protecting information collected from children under the age of 13.

- *The Disposal Rule of the Fair Credit Reporting Act (FCRA).*[3] The Fair Credit Reporting Act was created to regulate the consumer reporting industry and provide certain privacy rights in consumer

---

[1]  15 U.S.C. § 45.
[2]  16 C.F.R. §§ 312.1 to 312.13.
[3]  16 C.F.R. §§ 682.1 to 682.5.

reports. It regulates accurate and relevant data collection, limits the uses of credit reports, and provides individuals with the ability to access their information. With respect to cybersecurity, the FCRA sets the standards that businesses must meet if they collect consumer credit reporting information.

- *The Safeguards Rule of the Gramm-Leach-Bliley Act (GLBA).* Gramm-Leach-Bliley codified certain banking, securities, and insurance regulations to address concerns over how consumer data are collected, used, and shared. The GLBA sets security requirements on financial data. Companies that operate as financial institutions must provide administrative security (addressing the management of workforce risks and training), technical security (addressing computers, networks, encryption, and access controls), and physical security (addressing facilities, environmental safeguards, business continuity, and disaster recovery).

The following examples illustrate the FTC's role in enforcement and the types of cybersecurity matters that have led to enforcement actions.

*FTC v. Sandra Rennert.*[4] In this matter, an online pharmacy business allegedly violated the FTC Act by misleading consumers to believe that security controls were being used to protect their information when they had, in fact, not been implemented. The promises included encrypting personal information and using secure sockets layer (SSL) to secure transmitted data. An order was issued requiring the business to disclose the actual measures used to secure information, keep reasonable measures to protect personal information, and perform certain compliance monitoring and recordkeeping for five years.

*In re BJ's Wholesale Club, Inc.*[5] In this case, the warehouse club violated the Act by having insufficient data security which allowed unauthorized access to its networks, ultimately leading to payment card fraud. The FTC discovered that there were potential issues that involved the use of default user IDs and passwords, a lack of

---

[4] *FTC v. Sandra Rennert*, No. 992-3245, CV-S-00-0861-JBR (D. Nev. July 6, 2000).
[5] In re BJ's Wholesale Club, Inc., 140 F.T.C. 465 (2005).

encryption, issues with file access controls, and excessive payment card data retention. Allegations against BJ's in the complaint included failing to detect unauthorized access to networks and data, keeping data beyond the time allowed under banking regulations, and failing to protect consumer credit card data. BJ's was required to create a written information security policy (WISP) with appropriate administrative, technical, and operations measures addressed. The FTC also ordered a biannual security program assessment from an independent company and recordkeeping for 20 years.

In general, the FTC has acted when a company:

- Does not encrypt information while it was in transit or stored on in-store networks, especially when the company claims that they encrypt or are secure.

- Stores personally identifiable information permitting access to any person.

- Fails to use reasonable, readily available security measures to limit access, especially when it claims to do so.

- Fails to employ sufficient measures to detect unauthorized access.

- Fails to conduct security investigations when an incident occurs.

### Gramm-Leach-Bliley Act: Providing Guidance to Financial Institutions

The Gramm-Leach-Bliley Act,[6] or "GLBA," applies to financial institutions including banks, securities firms, insurance companies, and other businesses that provide financial services and products. The law regulates the collection, use, protection, and disclosure of nonpublic personal information, which means that the information is not publicly available and is capable of identifying a consumer or customer.

The GLBA specifically requires that financial institutions tell their customers about their information-sharing practices, give

---

[6] 15 U.S.C.A. §§ 6801 to 6809.

customers a right to opt out if they do not want their information shared with certain third parties (in accordance with the Financial Privacy Rule), and implement a WISP.[7] With respect to the security requirements, the WISP needs to incorporate reasonable technical, administrative, and physical safeguards to protect the security and confidentiality of customer information, protect against anticipated threats to the security and integrity of information, and protect against unauthorized access or use that could lead to substantial harm or inconvenience to the customer.

These requirements were heightened on December 9, 2021, when the FTC published a rule[8] amending the Safeguards Rule and requiring certain parts be put into writing and include specific criteria. The safeguards employed by the company must also address access controls, data inventory and classification, encryption, secure development practices, authentication, information disposal procedures, change management, testing, and incident response. The rule also exempted financial institutions that collect information on fewer than 5,000 consumers from creating a risk assessment, incident response plan, and annual reporting.

### Health Insurance Portability and Accountability Act (HIPAA) and Health Information Technology for Economic and Clinical Health Act (HITECH Act): Regulating Healthcare Providers, Health Plans, and Service Providers

The Health Insurance Portability and Accountability Act, commonly known as HIPAA, governs the duties of healthcare-related entities and service providers as they relate to personal health information.[9] HIPAA applies to "covered entities" including healthcare providers, health plans, and healthcare clearinghouses. The HIPAA is enforced by the Department of Health and Human Services, who has promulgated the following regulations:

- The Privacy Rule.[10] The Privacy Rule addresses the collection, use, and disclosure of protected health information (PHI). The Rule promotes strong privacy-based safeguards while still

---

[7] Safeguards Rule. 16 CFR part 314.
[8] 86 FR 70272-70314. 16 CFR 314.
[9] Pub. L. No. 104-191 (1996).
[10] 67 Fed. Reg. 53181 (August 14, 2002).

allowing individuals the right to access healthcare services and the data that these services create. The Rule requires appropriate safeguards to protect the privacy of PHI and sets limits and conditions on the uses and disclosures that may be made of such information without an individual's authorization. The Rule also gives individuals rights over their PHI, including rights to examine and obtain a copy of their health records, to direct a covered entity to transmit to a third party an electronic copy of their PHI in an electronic health record, and to request corrections.

- The Security Rule.[11] The Security Rule sets in place standards for the protection of the PHI. The Rule requires appropriate administrative, physical, and technical safeguards to ensure the confidentiality, integrity, and security of electronic PHI. The administrative safeguards include assigning responsibility for the security program to the appropriate individuals and requiring security training for employees. Physical safeguards include mechanisms to protect electronic systems and electronic PHI, such as limiting access to facilities to authorized individuals, protecting workstations for both onsite employees and teleworking employees, and protecting devices. Finally, technical safeguards include automated processes designed to protect data and control access, such as using authentication controls and encryption technology.

- The Data Breach Notification Rules.[12] The Notification Rules require notification of a breach of any unsecured PHI by a covered entity or its business associates. Under the Rules, a breach is an unauthorized acquisition, access, use, or disclosure of PHI that compromises the PHI privacy or security.[13] An acquisition, access, use, or disclosure of PHI is presumed to be a breach unless the company can show that there is a low probability that the PHI was compromised, based on a risk

---

[11] 45 C.F.R. § 164.310.
[12] Pub. L. No. 111-5 (2009); 74 Fed. Reg. 42740 (2009) (implementing the HITECH Act breach notification requirements).
[13] (45 C.F.R. § 164.402).

assessment that includes at least four factors. These factors include the nature and extent of PHI involved, the identity of the unauthorized person who used the PHI or to whom the disclosure was made, whether the PHI was actually acquired or viewed, and the extent to which the risk to PHI is mitigated.

- **The Transactions Rule.** The Standards for Electronic Transactions and Code Sets (the Transactions Rule) adopted standards for several transactions, including claims and encounter information, payment and remittance advice, and claims status. Any healthcare provider that conducts a standard transaction also must comply with the Privacy Rule.

## 9.2.2. Executive Orders

On February 12, 2013, President Barack Obama issued Executive Order 13636 titled *Improving Critical Infrastructure Cybersecurity.*[14] With the large number of intrusions into "critical infrastructure," President Obama sought to "enhance the security and resilience of the Nation's critical infrastructure and to maintain a cyber environment that encourages efficiency, innovation, and economic prosperity while promoting safety, security, business confidentiality, privacy, and civil liberties." The Executive Order defined critical infrastructure as "systems and assets, whether physical or virtual, so vital to the US that the incapacity or destruction of such systems and assets would have a debilitating impact on security, national economic security, national public health or safety." Further, the order requires federal agencies to develop and incentivize participation in a cybersecurity framework, sharing threat information received with the private sector. The Department of Homeland Security was tasked with leading and coordinating the regulatory efforts and identified 16 critical infrastructure sectors: chemical, dams, financial services, information technology, communications, defense, food and agriculture, nuclear, critical manufacturing, emergency services, government facilities, transportation, commercial facilities, energy, healthcare, and water.

---

[14] Improving Critical Infrastructure Cybersecurity, 78 FR 11739.

In May 2017, President Donald Trump also issued an Executive Order advancing cybersecurity within the federal government.[15] The Order focused on protecting critical infrastructure and training the workforce to mitigate and respond to attacks. Using the National Institute of Standards and Technology (NIST) Cybersecurity Framework, agencies were directed to provide a risk management report and to show a procurement preference for shared IT services.

President Joseph R. Biden took further executive action on May 12, 2021, with the Executive Order on *Improving the Nation's Cybersecurity*, encouraging more interaction between the government and private sector.[16] The Order calls for federal agencies to enhance the cybersecurity awareness and preparedness of the supply chain, notably software providers. Unlike the previous Executive Orders, President Biden set policy on cyber incident reporting obligations for federal contractors, which may create obligations for certain entrepreneurs.

**FIGURE 9.1**    Peters gets sued for negligent handling of PII.

Illustrated by Phillip Wandyez.

---

[15]  Executive Order 13800, Strengthening the Cybersecurity of Federal Networks and Critical Infrastructure.
[16]  Executive Order 14028.

*Somehow Peter always got selected for jury duty. But today, Peter was heading to court because one of his employees was suing him and the company. Peter's company was recently the target of a massive data breach that exposed the personally identifying information of all of his employees. His employee sued him for failing to protect the personal data, and the employee was a victim of identity theft as a result of the breach. Proper protocols could have prevented the harm. "I sure hope I have a good attorney" Peter thought. Of course, Peter somehow managed to find the worst attorney in town. He should have known when the sign on the attorney's office was misspelled "Atturney at Law."*

## 9.3. State Laws, Regulations, and Executive Orders

State laws tend to divide into two categories: (1) response to a data breach and (2) sector-related minimum standard laws.

### 9.3.1. Data Breach Laws

All 50 states and the US territories have enacted some level of data privacy and security laws.[17] Many of these laws require private or government entities to notify individuals of security breaches concerning personally identifiable information.[18]

---

[17] NATIONAL CONFERENCE OF STATE LEGISLATURES, http://www.ncsl.org/research/telecommunications-and-information-technology/security-breach-notification-laws.aspx

[18] NATIONAL CONFERENCE OF STATE LEGISLATURES, http://www.ncsl.org/research/telecommunications-and-information-technology/security-breach-notification-laws.aspx (last visited July 30, 2015). (The three states which have not enacted any such legislation include Alabama, New Mexico, and South Dakota).

Most of these laws contain similar provisions, which vary on a state-by-state basis. These topics include:

- *The definition of Personal Information or Personal Identifying Information.* Many states consider a person's first name or initial, a last name, and either (1) a social security number, (2) driver's license or state identification, or (3) an account number with a password to be the hallmarks of personal information. Definitions usually exclude any information that can be found in a publicly available source like a phonebook or in widely distributed media.

- *Who is covered by the law (i.e., which business types).* Generally, state laws for data breach apply to those who conduct business within the state. Some states are narrower and only include those who work as data brokers.

- *The definition of a "security incident" or "breach."* The definitions of breach vary from state to state and range from "unauthorized access to or acquisition of electronic files, media, databases or computerized data containing personal information, when access to the personal information has not been secured by encryption or by any other method or technology that renders the personal information unreadable or unusable" in Connecticut to an event that causes or likely to cause identity theft or other material harm in Kansas and South Carolina.[19]

- *What factors are required for notification?* Notification requires an analysis that is highly dependent on state law. Each state has its own set of factors, but these often include an assessment of the number of individuals impacted, their location, and the type of information (e.g., was it personal health information subject to HIPAA).

- *Who has to be notified?* Typically, the residents of a state that are impacted by a data breach would be notified, if a state law requires notification. Certain states, however, create different and farther-reaching strategies. In Texas, for example, in

---

[19] Conn. Gen. Stat. 36a-701b; Kan. Stat. 50-7a01; and S.C. Code 39-1-90.

addition to Texas residents, a company must notify residents of states who do not have a reporting law. In some states, the state attorney general's office must be notified if a certain number of individuals are impacted. For example, North Dakota and Oregon have reporting thresholds of 250 individuals where South Carolina and Virginia have reporting only if more than one thousand individuals are impacted. Additionally, some states require the notification of nationwide credit reporting agencies or other state governmental agencies like insurance commissioners. Some states also encourage or mandate reporting to law enforcement.

- *When do individuals need to be notified?* State laws also contemplate timing for notification. Most states require a company experiencing a breach to report in "the most expeditious time possible and without unreasonable delay." Other states create a limit that ranges from 24 h to 45 days. These potentially tight timeframes and requirements dictate action and a prompt response once a breach is found.

- *What must be included in the notification letter?* Similarly, each state may require certain information to be contained in a notification letter going out to state residents. Generally, states require a description of the breach, a description of the type of data impacted, what the business is doing to prevent this from occurring again, a phone number where the individual can obtain more information, advice on getting free credit reports and how to review their accounts, contact information for the major consumer reporting agencies, and contact information for the FTC.

- *Any exceptions to reporting.* Although reporting is the typical response to a breach that impacts a number of individuals, there are times when state-mandated reporting is not required. First, if companies are required to report under a federal law (HIPAA or GLBA), reporting at a state level may not be required. Indeed, with federal reporting, it is likely that the same information will be passed along to the impacted individuals, and state-level reporting would be merely redundant. A second exemption in some states applies if the company follows its own breach notification procedures as long as they are compliant with the law. Finally, some states have safe harbors for information that is encrypted, redacted, unreadable or unusable. This information has been deemed to be less risky and generally there either has not been a compromise, or if there is, there is no harm.

- *The penalties, fines, and possibility of a private right of action (where a citizen can sue the company).* Many states allow enforcement through the respective state attorney general. Some companies, in certain situations, could be subject to penalties. As an example, Missouri allows the attorney general to bring an action to "seek a civil penalty not to exceed one hundred fifty thousand per breach of the security of a system or series of breaches of a similar nature that are discovered in a single investigation."[20]

## 9.3.2. Minimum Standard and Reasonable Data Security Measure Laws

Beyond data breach statutes, some states have enacted laws to be more proactive with respect to cybersecurity.

There are state laws that require organizations to maintain "reasonable data security measures" to protect personal information from unauthorized access, acquisition, destruction, disclosure, modification, and use.[21] Other states have more specific data security obligations including:

- Alabama[22] requires organizations to consider certain factors when implementing and maintaining reasonable security measures, based on the data type and the organization's business activities.

---

[20] Mo. Rev. Stat. 407.1500.
[21] Alabama, Arkansas, California, Colorado, Connecticut, Delaware, DC, Florida, Illinois, Indiana, Kansas, Louisiana, Maryland, Nebraska, Nevada, New Mexico, New York, Oregon, Rhode Island, Texas, Utah and Virginia.
[22] Ala. Code 1975, § 8-38-3.

- California[23] requires connected device manufacturers to take steps to ensure the security of devices and the information they contain.

- Massachusetts[24] holds some of the most stringent information security program requirements in the country, including requiring the development, implementation, and maintenance of a comprehensive WISP.

- Nevada[25] imposes additional requirements, including data encryption, under certain circumstances based on whether an organization accepts payment card transactions.

- New York[26] requires reasonable safeguards where a company either complies with another recognized data security regime or has certain elements of security implemented.

- Oregon[27] requires companies to maintain reasonable data security measures, including specific administrative, physical, and technical safeguards, and connected device manufacturers to equip devices with reasonable security features.

- Rhode Island[28] requires entities to implement and maintain a scalable risk-based information security program.

Of these laws, there are several that are commonly considered to be either comprehensive or detailed for the sector in which they operate.

### New York Department of Financial Services (NYDFS) Cybersecurity Requirements for Financial Services Companies[29]

The State of New York implemented financial regulations requiring certain companies subject to banking, insurance, or financial services law to certify that their data security is sufficient by

---

23  Cal. Civ. Code § 1798.91.04.

24  M.G.L. ch. 93H, § 2; 201 Mass. Code Regs. 17.01-05.

25  NRS 603A.210(2).

26  N.Y. Gen. Bus. Law § 899-bb(2).

27  Or. Rev. Stat. § 646A.622.

28  R.I. Gen. Laws § 11-49.3-2.

29  23 NYCRR §§ 500.0 to 500.23.

identifying cybersecurity threats and employing a defense infrastructure to protect against threats, to use a system that detects intrusions, and to respond and recover from any incidents. These prescriptive regulations are among the most aggressive in the nation as they require businesses to file a certificate of compliance annually. Certain companies who either have fewer than ten employees located in New York, less than $5 million in gross revenue in the last three fiscal years from their own and their affiliates' NY business operations, or less than $10 million in total year-end assets are exempt from the law and are only required to file a notice of exemption.

The obligations for companies under the NYDFS Cybersecurity law include creating and maintaining written policies and procedures, performing risk assessments, and having certain core functions and safeguards as directed by the law. Those needing to conform to this law find that the requirements can be challenging, and companies often seek the assistance of third-party service providers to assist in creating the necessary framework within the company.

### Massachusetts State Security Law and Regulations[30]

Unlike most state regulations or laws, Massachusetts requires covered businesses to adopt a WISP with specific security measures. It also has broad reach and is applicable to companies regardless of where they are, as long as they own or license the information of a state resident. Fortunately, the WISP can be appropriate to the size, scope, and type of business along with the amount of data, need for security, and resources available to accomplish the task. The specific requirements that must be addressed include: Designating a program oversight person; Identifying and assessing reasonably foreseeable risks to security; Evaluating and improving the effectiveness of existing safeguards; Developing policies; Imposing discipline for policy violations; Assuring that terminated employees cannot access records; Overseeing service providers; Implementing physical access restrictions; Monitoring

---

[30]   201 Code Mass Regs. 17.01 to 17.05.

performance; Upgrading as necessary; Reviewing the plan annually; Documenting performance.

### California Internet of Things Law

In 2018, California became the first state to pass a law addressing cybersecurity in the Internet of Things (IoT).[31] The purpose of the law is to require manufacturers of IoT devices to provide "reasonable security features" that protect user privacy. Specifically, any IoT device must be equipped with appropriate features that protect information contained in the device from unauthorized access, destruction, use, modification, or disclosure. In 2022, the law was amended to allow compliance with the NIST criteria and labeling to constitute compliance with the IoT law.

## 9.4. European Union and International Requirements

Depending on the nature of the enterprise, international requirements for data security may apply.

In the European Union (EU), the General Data Protection Regulation (GDPR) is a legal framework present in the EU which allows EU citizens additional controls over their personal information. The regulations create further requirements and terms for those who gather, collect, use, or manage such data. Under the GDPR, the collection of personal information from EU citizens must be:

- *Fair and transparent*: Processed lawfully, fairly, and in a transparent manner in relation to individuals.

- *Legitimate and explicit*: Collected for specified, explicit, and legitimate purposes and not further processed in a manner that is incompatible with those purposes; further processing for archiving purposes in the public interest, scientific or historical research purposes, or statistical purposes shall not be considered to be incompatible with the initial purposes.

---

[31] California Legislative Information. (2018). Assembly Bill No. 1906. leginfo.legislature.ca.gov/faces/bill-TextClient.xhtml?bill_id=201720180AB1906

- *Adequate and limited*: Adequate, relevant, and limited to what is necessary in relation to the purposes for which they are processed.

- *Accurate and current*: Accurate and, where necessary, kept up to date; every reasonable step must be taken to ensure that personal data that are inaccurate, having regard to the purposes for which they are processed, are erased, or rectified without delay.

- *Time limited*: Kept in a form which permits identification of data subjects for no longer than is necessary for the purposes for which the personal data are processed; personal data may be stored for longer periods insofar as the personal data will be processed solely for archiving purposes in the public interest, scientific or historical research purposes, or statistical purposes subject to implementation of the appropriate technical and organizational measures required by the GDPR in order to safeguard the rights and freedoms of individuals.

- *Secure and protected*: Processed in a manner that ensures appropriate security of the personal data, including protection against unauthorized or unlawful processing and against accidental loss, destruction, or damage, using appropriate technical or organizational measures.

Citizens in the EU are also given specific rights with respect to their data including the right to be informed, the right to access, the right of rectification, the right to erasure, the right to restrict processing, the right to data portability, the right to object to uses, and the right to safeguards regarding automated decision-making and profiling.

Other international requirements include:

- Brazil: The Lei Geral de Proteçao de Dados Pessoais (LGPD) was modeled directly after the GDPR and is nearly identical in terms of scope and applicability, but with less harsh financial penalties for noncompliance.

- South Korea: South Korea's Personal Information Protection Act, 개인정보 보호법, includes many GDPR-like provisions such as requirements for gaining consent, the scope of applicable data, appointment of a Chief Privacy Officer, and limitation and justification of data retention periods.

- China: China has many comprehensive laws in the privacy and cybersecurity realm that are currently undergoing regulatory clarifications. For those operating within China, it is important to consult with qualified legal counsel to understand all applicable requirements.

# 9.5. Practical Considerations: How Cyber Law Can Directly Impact Your Business

## 9.5.1. General Recommendations

The FTC, in many different guidance documents, presents a strategy for mitigating risk, including the potential of legal claims. In a nutshell, companies should:

- Provide reasonable security for data collected that is scaled to the business. This means understanding and adapting to the level of data sensitivity (asking "is it sensitive personal identifying information," for example).

- Protect data according to the requirements and standards of the industry. For those in the automotive sector, adherence (as appropriate) to SAE standards would be critical.

- Understand the type of risk and threat actors seeking to attack that type of business.

- Follow any reasonable industry standards and adopt reasonable protections as they are available.

- Maintain and retain data only for as long as needed to perform the business task.

- Limit sharing with any third-party source unless under contract and as needed to perform the business function.

Other key parts of any information security program that will assist in compliance with laws include ongoing risk identification and assessment, risk management, policies, technical safeguards, training, vendor and supply chain management, monitoring, and incident response.

### 9.5.2. Cybersecurity Terms in Contracts

Cyber terms and conditions are more and more frequently appearing in general contracts. The obligations include: Having protections in place to mitigate the risk in collecting and using the personal information of customers, employees, or others; Assuring the protection of customer's trade secrets or other confidential or proprietary information; Adhering to industry standards when accepting certain forms of payment, including credit cards, other payment cards, and direct payments from bank accounts; Insuring against cyber risk; Demonstrating compliance with generally accepted industry standards for various other business purposes, including contracting with downstream vendors.

It is important for any company to review their customer's terms and conditions and understand the cyber requirements present. It is also good practice to place cyber terms in any downstream agreements with vendors or suppliers to uphold the same relationship.

## 9.6. Conclusion

As an entrepreneur, the legal landscape may feel daunting, especially where operations are based upon the collection, use, or transfer of data related to individuals. From a legal perspective, it is critical to identify where you are doing business. Are you operating offices across the US? Do you have or intend to have locations and employees globally? And are you targeting, selling, or offering your products or services in certain locations around the globe that may put cybersecurity laws on the table? And do you collect information about customers or clients in various locations?

Once you understand the scope of your operations, the data you collect and use, and where you are targeting your efforts, it is important to know any industry specific rules. As noted above, certain sectors are governed by more specific laws. Finally, consider how policies, procedures, and plans can help you meet legal compliance. These policies can range from privacy policy and terms of use for a website to breach notification plans from employee handbooks to business interruption plans. Attention to these details can help avoid difficult challenges further down the road.

# 10

# Cyber Economics: How Much to Spend on Cybersecurity

C. Ariel Pinto and Luna Magpili

## 10.1. Introduction

Earlier chapters described how to secure your communications, financial transactions, data, Internet of Things, and—more importantly for entrepreneurs—trade secrets. However, as most entrepreneurs also realize, there could only be so much to spend on all of these. We often think of economics in terms of money—the cost of products and services. However, as it pertains to cybersecurity for an entrepreneur, the better perspective of economics is looking at all the resources (including, but not purely, money) and efforts spent on anything and everything "cybersecurity." This includes that important balance between the resources spent and the benefits the company gains in return. For an entrepreneur, those benefits may come in various forms that add value to the company's products and services.

As an example, consider a new laptop computer that already comes with built-in fingerprint reader technology. As a potential buyer of this product, that technology may be "valued" for its added authentication security and convenience. However, that technology places additional costs to manufacture that laptop, but presumably, it is economically justified given that it may make the product more appealing to buyers. Or consider the reality of remote work where a company requires all computers to be configured in a specific way before they can be used by employees for their work. This added security brought by managed configuration may make the company's trade secrets more secure but would also need Information Technology (IT) personnel to spend time configuring each computer and possibly needing to purchase and install specific software and applications (apps) to do this. How does the company's cybersecurity officer (as an entrepreneur, that's you!) justify these types of efforts and costs with the benefits they may bring?

## 10.2. Value of Your Product or Service

All entrepreneurs recognize that we secure our cyber resources not just for the heck of it. Rather, we do it because it has become a necessary part of doing business.

First, let us think of the value of your product or service. This refers to the benefits that the product or service offers to the target customers and how much the customer may be willing to pay for it. So it is important to keep in mind the customer and the importance of cybersecurity as it relates to the product and service being offered to them. What we know is that the overall value of a product or service is the degree to which it meets or exceeds customers' expectations. We have heard this in the past from other entrepreneurs that cybersecurity may not really be in the minds of their target customers. Indeed, products like pastries and clothes may seem to have nothing to do with cybersecurity. But later, we will realize that because cyber activities are so embedded in almost every process in a company—yes, think of an automated baking machine, the procurement process for the raw materials, and the payment process for customers—cybersecurity permeates ordinary products and services far more than it may meet the eye.

But how do we "add" value to products and services during their design and build? Primarily in two ways:

- Manufacturing of products

- Delivery of service

These two value-adding activities together greatly influence pricing strategy and brand messaging. So why is it important to ask how much to spend on cybersecurity? Well, just like any investment, a successful manager-entrepreneur would like to know if it is "worth it." The "worth" of investment in cybersecurity needs to be viewed from at least two perspectives: as a cost center and as a profit center.

**FIGURE 10.1**  Peter has no money to spend.

Illustrated by Phillip Wandyez.

*As Peter sat in his office, he thought to himself,*
*"This cyber stuff might actually be important*
*after all... Maybe I should put some money*
*toward cybersecurity for my startup... Oh wait,*
*I do not have any money!" Peter in fact did not*
*have any money. Between his pricey lawsuit*
*(that he lost) and his other failed business*
*ventures, Peter could barely afford to keep the*
*lights on. "Besides, where would I even start?"*
*Peter mused. "How would I know what cyber*
*protections give me the best return on invest-*
*ment? If only someone would write a book*
*about this stuff!"*

## 10.3. Cybersecurity as a Cost Center versus a Profit Center

A cost center is typically a department or function within a company that does not directly add to profit but still costs the organization money to operate. Rather, it ensures the proper functioning of the key profit-generating units of the business, and in that process, it incurs necessary costs. Examples of cost centers are departments such as accounting, human resources, maintenance, research and development (R&D), and—most relevant to this chapter—Information Technology (IT). These are all necessary functions but do not directly add to the profit of each product sold and service delivered. All of these departments nowadays need some form of cybersecurity, e.g., accounting, intellectual property, and employee and customer data need to be protected from hackers; IT needs malware protection and a business-continuity plan; R&D data need to be protected from corporate espionage. Hence, cybersecurity spending that does not directly add to profit but still costs the business money to operate can be considered a cost center.

On the other hand, a profit center refers to a department or function within a company which directly adds to the bottom line or profits of every product sold or service delivered. Examples of profit centers are departments in a company such as purchasing, production, warehousing, marketing, and sales.

For example, automobiles connected to the cyber-infrastructure are a common thing—GPS, hotspots, OnStar, smart cars, and, of course, semi- and fully-autonomous vehicles. All the engineering and manufacturing costs that come with securing all these technologies (i.e., cybersecurity built into the product) may add profit to each car sold with buyers who see the cybersecurity feature as valuable and hence will pay a premium for them.

On the other hand, cybersecurity can also be added as a pure service on top of more traditional services, i.e., data protection offered by cloud service providers for customers who store their data in the cloud, confidentiality protection offered by financial institutions for their customer's personal data, etc. Like that of the connected automobile, all the costs that come with these additional cybersecurity services (i.e., cybersecurity built into the basic service) directly contribute toward profit from customers who see these features as valuable and hence will pay a premium for them.

Furthermore, profit centers can also be entire operational units such as separate branches of a retail store or a facility in a distinct geographic location. Consider Industry 4.0 which would require technologies such as sensors, robotics, and instrumentations for controlling production machines. Cybersecurity is necessary to harness the benefit of Industry 4.0, and hence, it would be considered as adding to the bottom line or profits of the company, i.e., profit center. These technologies are also referred to as Operational Technologies (OT) and can be distinguished from IT in that IT refers to computers and other equipment used to support, but not directly affect, operations, while OT directly affects operations. Hence, when cybersecurity directly adds to the value of a product or service, we can relate cybersecurity spending to a profit center and would expect tangible profit from such spending. The distinction between cost and profit centers can be summarized in Table 10.1.

**TABLE 10.1** Summary of profit centers and cost centers.

| | Cybersecurity as profit center | Cybersecurity as cost center |
|---|---|---|
| **Primary purpose** | To add value to products or services; to increase profit | To secure processes and functions supporting primary operations and key profit-generating units of the business |
| **Primary criterion for evaluation** | Increase in sales or profit from product or service | Reduction in cyber risk of supporting primary operations and key profit-generating units of the business |
| **Examples** | Multifactor authentication that comes with online services offered by banks to its clients | Multifactor authentication for bank employees to access the bank network and serve clients |
| | Encryption for onboard Wi-Fi capabilities on cars for its driver and passengers | Encryption for Wi-Fi connection used by assembly robots in a car factory |

# 10.4. How Much to Spend: Common Economic Measures of Cybersecurity Spending

## 10.4.1. Return on Investment (ROI)

One of the most basic and common ways to figure out how much to spend on an investment is by calculating the return on investment (ROI). ROI is a performance measure most used to evaluate an investment or compare several different investments. It tries to directly measure the amount of return on a particular investment, relative to the investment cost. ROI can also be used in economically comparing the cost and benefits of cybersecurity spending. The basic formula for ROI is:

$$\text{ROI} = \frac{\text{Expected returns} - \text{Cost of investment}}{\text{Cost of investment}} \qquad (10.1)$$

*Example 1:* As an example, Peter is considering two alternative authentication technologies for company-wide implementation. Shown below is the relevant information for Technology A and Technology B, including the resulting ROI (Table 10.2):

**TABLE 10.2**  Data for ROI calculation in Example 1.

|  | Technology A | Technology B |
|---|---|---|
| **Expected benefit** | $1,000,000 | $1,500,000 |
| **Expected cost** | $500,000 | $600,000 |
| **ROI** | 100% | 150% |

ROI for Technology A:

$$\text{ROI} = \frac{\$1,000,000 - \$500,000}{\$500,000} = 100\% \qquad (10.2)$$

ROI for Technology B:

$$\text{ROI} = \frac{\$1,500,000 - \$600,000}{\$600,000} = 150\% \qquad (10.3)$$

Numerically, the ROI for Technology B is higher than for Technology A. Peter can interpret this as "for every dollar spent in B, it provides a dollar-and-a-half of benefit. While for Technology A, the return would only have been a dollar." Hence, an entrepreneur may use ROI to justify choosing Technology B over Technology A because it would give a higher ROI, dollar per dollar.

*Example 2:* Another example could be if Peter is trying to figure out if his company should offer its customers a "one-year free" credit monitoring service when they sign up as a new customer. This "free'" service to both current and new customers would cost his company an estimated $200K. However, the buzz that this "free" service would generate in the marketplace would generate an expected additional $300K of revenue. This information, as well as the resulting ROI, is shown in Table 10.3.

$$\text{ROI} = \frac{\$300,000 - \$200,000}{\$200,000} = 50\% \qquad (10.4)$$

**TABLE 10.3** Data for ROI calculation in Example 2.

| | |
|---|---|
| **Expected benefit** | $300,000 |
| **Expected cost** | $200,000 |
| **ROI** | 50% |

The calculated ROI of 50% can be interpreted as "for every dollar spent in this added service, it provides a half-a-dollar of benefits." This ROI may turn out to be not good enough to pursue. Hence, Peter may just decide not to offer such "free" service to his customers.

### 10.4.2. Risk-Based Return on Investment for Cost Center Spending

Simply, risk-based return on investment (RROI) is the ratio, or percentage, between the net benefit of implementing a risk mitigation solution and their implementation costs, akin to ROI. However, it guides entrepreneurs to more meaningful answers to the following questions than typical ROI may provide:

- What are my options in cybersecurity investments?
- Which cybersecurity risk am I concerned about?
- How well are the options protecting me from these risks?
- What value am I adding to my products and services?
- How much should I invest?
- Have I invested enough?

It is apparent that, unlike typical investments of offering an added service to generate more sales (a profit center), there are cases when the expected benefit of spending on cybersecurity is a reduction in the expected loss from risk, i.e., cybersecurity in IT as a cost center. The fundamental difference this makes in economics is the assumption of reinvestment, that is, the "benefit" of reduction in expected loss does not necessarily translate into additional resources, which companies would typically use for other productive endeavors. So investing in cybersecurity in your IT system as a cost center does not translate into producing more products. Furthermore, cybersecurity spending is usually a composite of many technologies and policy-related investments (e.g., multifactor authentication technology and employee training). Hence, there may be instances when the traditional ROI needs to be refined to capture these details unique to cybersecurity spending—in the form of RROI, Return on Security Investment (ROSI), and other similar measures. This section will describe these suites of measures by looking specifically at RROI. The formula for RROI is:

$$RROI = \frac{\text{Baseline scenario} - \text{Residual risk} - \text{Cost}}{\text{Cost}} \qquad (10.5)$$

*Baseline Scenario:* The grand total of exposure a company may have to cybersecurity losses if it had no risk mitigation in place. This exposure could be comprised of various types which are of concern to a specific company, e.g., ransomware and DDoS. For example, a financial institution may be concerned mostly about hackers that can steal customers' information, while an online retail company may be more concerned with DDoS which may cripple their public-facing website.

*Residual Risk:* The expected value of consequences with risk mitigations in place as a function of its effectiveness. Like the baseline scenarios, a residual risk can still be comprised of the same types even after the implementation of corresponding cybersecurity solutions but with lower values.

*Cost:* The cost of all cybersecurity investments being considered, like that in the earlier ROI.

The general analysis framework for RROI has three steps:

1. Estimate the effectiveness (i.e., net bypass rate) for all security solutions.
2. Calculate incident risk and baseline scenario.
3. Calculate net benefits and, ultimately, RROI.

***Example 3:*** For example, consider that Peter is evaluating five technologies currently being implemented for mitigating four types of cybersecurity incidents. For one year, Peter recorded network security incidents classified into the four types and the associated damages including the cost of resources used to repair damaged information as well as any lost productivity. This information is summarized in Table 10.4.

**TABLE 10.4** Occurrences and damage by incident type.

| | | Incident type | | |
|---|---|---|---|---|
| **Characteristics** | **DDoS** | **Ransomware attack** | **Data exfiltration** | **Root compromises** |
| **Observed occurrences** | 4 | 5 | 3 | 6 |
| **Observed damages ($)** | $3,000 | $5,000 | $10,000 | $4,000 |
| **Observed damages, total ($)** | $22,000 | | | |

***Step 1: Estimate the net bypass rate for all security solutions.*** Peter knows that these are only the tip of the iceberg and that there could be more close-call incidents which the existing security system were able to successfully prevent; hence, no damage is recorded. After close collaboration with vendors and IT professionals, Peter estimated how effective (conversely, how ineffective) the security solutions were in preventing incidents, as shown in the top part of Table 10.5.

**TABLE 10.5** Bypass rates, observed damage, and incident risk.

| Security solution (vs bypass rates) | DDoS | Ransomware attack | Data exfiltration | Root compromises |
|---|---|---|---|---|
| Multifactor authentication | 0.8 | 1 | 0.8 | 0.8 |
| Vulnerability patching program | 0.4 | 1 | 0.4 | 0.4 |
| Intrusion prevention and detection | 0.15 | 0.15 | 0.15 | 0.15 |
| Configuration management system | 0.6 | 0.8 | 0.5 | 0.55 |
| Training program | 0.8 | 0.8 | 0.8 | 0.9 |
| Net bypass rate | 0.0230 | 0.0960 | 0.0192 | 0.0238 |
| Observed damage ($) | $3,000 | $5,000 | $10,000 | $4,000 |
| Incident risk | $130,200 | $52,100 | $520,800 | $168,400 |
| Baseline scenario | $871,500 | | | |

Part of Peter's role is to establish how effective (or ineffective) cybersecurity solutions are to various types of risks. This insight is captured by the Bypass Rate, which is the effectiveness of the risk mitigation solution, i.e., 0% = perfectly effective in preventing a certain type of cyber incident and 100% = totally ineffective.

Peter also knows that it is common for a certain cybersecurity solution to address more than one type of incident. For example, multifactor authentication partially addresses the spread of malware as well as unauthorized access to company data, in the same manner that a virus scanning app may prevent spread of malware but may not prevent unauthorized access. Hence, Net Bypass Rate may be used to describe the exposure of the entire company to a specific type of cybersecurity incident after all security solutions have been considered.

$$\text{Net Bypass Rate}\left(\text{Incident type}\right)$$
$$= \prod \text{Bypass Rate}\left(\text{Incident type, security solution}\right) \tag{10.6}$$

*Step 2: Calculate incident risk and baseline scenario.* The incident risk and baseline scenario are calculated as follows:

$$\text{Incident risk} = \frac{\text{Observed damage}\left(\text{Incident type}\right)}{\text{Net Bypass Rate}\left(\text{Incident type}\right)} \tag{10.7}$$

$$\text{Baseline scenario} = \sum_{\text{for all Incident types}} \text{Incident risk}\left(\text{Incident type}\right) \tag{10.8}$$

Keep in mind that this is the grand total of exposure a company may have to cybersecurity losses if it had no cybersecurity solutions in place.

*Step 3: Calculate net benefits and ultimately, RROI.* Hence:

$$\text{Residual risk}\left(\text{Security solution}\right) = \sum \Big[ \text{Incident risk}\left(\text{Incident type}\right)$$
$$\times \text{Bypass rate}\left(\text{Security solution}\right) \Big] \tag{10.9}$$

Noticeably, "Baseline Scenario – Residual Risk" in RROI is like "Expected Return" in the earlier ROI. However, with RROI, an entrepreneur can now assemble the notion of "return" of cybersecurity spending—the Net Benefit, from its essential components: Baseline Scenario and Residual Risk, which are more readily available. For this example, the RROI for the various cybersecurity solutions are shown in Table 10.6.

**TABLE 10.6** RROI of each individual security solution acting alone.

| Security solution | Residual risk ($) | Net benefit ($) | Cost | RROI |
|---|---|---|---|---|
| Multifactor authentication | $707,600 | $122,900 | $41,000 | 300% |
| Vulnerability patching program | $379,900 | $333,600 | $158,000 | 211% |
| Intrusion prevention and detection | $130,700 | $64,800 | $676,000 | 10% |
| Configuration management system | $472,800 | $11,700 | $387,000 | 3% |
| Training program | $714,000 | $(161,500) | $319,000 | −51% |

Peter can now compare the RROI for each security solution. For example, every dollar spent in multifactor authentication protects the business $3 worth of potential loss. Correspondingly, these are $2.11, $0.1, and $0.03 for vulnerability patching program, intrusion prevention and detection, and configuration management system, respectively. Every dollar invested provides a corresponding reduction of potential loss from different security risks. The training program, however, gives a negative RROI, which means the cost of training does not give enough benefit of protection across the various security risks.

## 10.4.3. Delayed Net Present Value and Catastrophic Cybersecurity Incidents

But certainly, not all cybersecurity incidents are the same. There are those that seem to be nightmarish and may keep entrepreneurs like Peter awake at night, e.g., a major ransomware attack resulting in the total shutdown of operations and malware that cripples both IT and OT systems for days or weeks. Even then, there will never be enough resources to spend to prevent these catastrophic cybersecurity incidents. Much like how we cannot prevent a hurricane or an earthquake, no amount of resource spending can totally prevent these types of incidents from happening. So for this scenario, the recourse would be to prepare a company so it can recover quickly from such incidents when they happen. Hence, there is this omnipresent dilemma in which, to paraphrase Benjamin Franklin's famous proverb, an ounce of prevention is worth a pound of cure. There is a tradeoff between loss prevention and loss reduction. This leads an entrepreneur to look at cybersecurity spending beyond just the assurance that the company survives a major cyber incident, but also how fast it can recover given the time value of its resources. Hence, cybersecurity spending may be thought of through a planning horizon of time.

Time consideration of investments is usually calculated based on the Net Present Value (NPV) and its various extensions. NPV is calculated for the investment at the appropriate discount rate, and then the cost of the investment is subtracted, summarized in the formula:

$$\text{NPV} = -C_0 + \frac{C_1}{1+r} + \frac{C_2}{(1+r)^2} + \cdots + \frac{C_T}{(1+r)^T} \qquad (10.10)$$

where
   $-C_0$ is the initial investment
   $C$ is the cash flow
   $r$ is the discount rate
   $T$ is the time

This provides insight about whether the investment should be made, that is, if the NPV is positive. The evolving cybersecurity landscape has shown that if we do not do anything to prevent incidents from occurring, the tendency would always be for them to occur sooner rather than later—particularly for incidents caused by malicious agents like hackers. Hence, one can think of cybersecurity spending as a loss control strategy that pushes the occurrence of catastrophic cybersecurity incidents further into the future—thus the term "Delayed" NPV, not that it will never happen. For an entrepreneur looking at cybersecurity spending, this Delayed NPV may be negative because some events may never be totally prevented. But still resources can be thought of in terms of their effects on "pushing" a catastrophic cyber incident from occurring into the future, as well as the reduction of its consequences if it does occur. This is a strong reminder for entrepreneurs of the unfortunate truth that managing cybersecurity is a never-ending process. But at the same time, cybersecurity spending may also serve as loss reduction strategy that reduces the consequence of incidents if they do occur.

The general analysis framework for Delayed NPV has the following steps:

1. Estimation of time to event of catastrophic cybersecurity incident.
2. Estimation of the consequences if the event does occur.

3. Identification and description of prevention and loss reduction alternatives.

4. Delayed NPV calculation.

These steps will be described in the form of a two-part example.

***Example 4, Part 1: First-Generation Artificial Intelligence (AI) Authentication App.*** Peter is planning to deploy a new AI authentication app worth $5M for critically important employees to prevent the entry of malware into the most secure part of the company's IT/OT system. It was estimated that if this catastrophic cyber incident occurs, it will cost Peter's company a staggering $300M in estimated damages. This event is believed likely to occur as early as five years from now if no intervention is done. Like other investments, the option to deploy this new AI authentication app needs to be compared with the do-nothing option. For illustration, consider a prevailing discount rate of 5%. Relevant information is shown in Table 10.7.

**TABLE 10.7** Data on do-nothing and first-generation AI authentication app alternatives.

|  | Do-nothing | First-generation AI authentication app |
|---|---|---|
| Initial cost | 0 | $(35,000,000) |
| Yearly cost | 0 | $(5,000,000) |
| Consequences | $(300,000,000) | $(300,000,000) |
| Discount rate | 5% | 5% |

How effective should this AI authentication app be in preventing such an event to make it cost-justified?

*Steps 1 and 2:* Estimation of time to event of catastrophic cybersecurity incident and estimation of the consequences if the event does occur. From the narrative of this example, the time to event (*T*) is five years and consequences (*C*) is $300M, respectively.

*Step 3: Identification and description of prevention and loss reduction alternatives.* Simply, the alternatives are to either implement AI authentication app or do nothing.

*Step 4: Delayed NPV calculation.* Using the NPV formula for various values of time ($T$), Peter generated the Delayed NPV values for both alternatives, as shown in the following graph. Peter realizes that the First-generation AI authentication app needs to be effective enough to delay the catastrophic event by more than 14 years for it to be better than do nothing, as shown in **Figure 10.2**. Now, it will depend on Peter's evaluation if indeed this technology can deliver such benefit.

**FIGURE 10.2**   NPV versus time ($T$) graph for first-generation AI authentication app and for the do-nothing options.

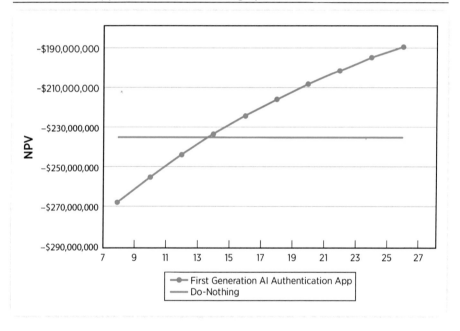

***Example 4, Part 2: Second-Generation AI Authentication App.*** Consider the preceding example, but say that, since the last analysis, there has been considerable improvement in the second generation of this AI technology coupled with policies on how information coming from this technology can be used by the company in its business continuity plans. As a result, the technology not only delays the occurrence of the event but also contributes to lower the damages of common events at a yearly basis by $4.5M. Furthermore, if the catastrophic event occurs, the damages will be much lower, from \$300M with the first generation down to \$250M for the second generation of the technology. However, this additional benefit comes at a higher cost of annual operation, from \$5M up to \$10M. Hence, the yearly net benefit of this newer technology is \$4.5M − \$10M = −\$5.5M. Like the first case, consider a prevailing discount rate of 5% for purpose of illustration. This information is summarized in Table 10.8.

**TABLE 10.8** Data on do-nothing and second-generation AI authentication app alternatives.

|  | Do nothing | Second-generation AI authentication app |
|---|---|---|
| Initial cost | 0 | $(35,000,000) |
| Yearly cost | 0 | $(5,500,000.00) |
| C | $(300,000,000.00) | $(250,000,000.00) |
| Discount Rate | 5% | 5% |

How effective should this second-generation AI authentication app be in preventing such an event to make it cost-justified?

*Steps 1 and 2: Estimation of time to event of catastrophic cybersecurity incident and estimation of the consequences if the event does occur.* From the narrative of this example, the time to event (*T*) is five years and consequences (*C*) is \$300M, respectively.

*Step 3: Identification and description of prevention and loss reduction alternatives.* Simply, the alternatives are to either implement second-generation AI authentication app or do nothing.

*Step 4: Delayed NPV calculation.* Using the Delayed NPV formula for various values of time (*T*), Peter generated the values for both alternatives as shown in the following graph. Peter realizes that the second-generation AI authentication app needs to be effective enough to delay the catastrophic event by about 8.25 years for it to be better than do nothing (Figure 10.3). Now, it will depend on Peter's evaluation if indeed this technology can deliver such a benefit.

Through the notion of Delayed NPV, Peter was able to bring together details of costs and benefits of cybersecurity not at a single point in time, but rather through a planning horizon. Furthermore,

Peter was also able to fold into the analysis the current nature of cybersecurity spending as a continuous endeavor that may not totally prevent occurrences of cyber incidents but rather the limited shelf-life of mitigation strategies.

**FIGURE 10.3**  NPV versus time (*T*) graph for second-generation AI authentication app and for the do-nothing options.

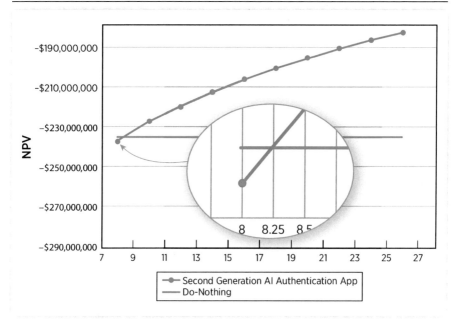

## 10.5. Estimating Costs, Benefits, and Other Information

There is a lot of estimating involved in the calculation of ROI, RROI, and NPV. For costs, it can be easier because we usually have some data on the costs of cybersecurity technology, tool, or app. However, determining the dollar benefits of using a cybersecurity tool that has the potential to increase market sales, or estimating the loss of a cyberattack if it is not prevented by using a particular cybersecurity technology, is honestly mostly a lot of guesswork. However, these types of analyses are still significantly better than depending on a crystal ball and a prayer that nothing bad happens. So how do we make our guesses good estimates?

First, we should use whatever data or information we have. Nothing beats real data. If we can base our costs or losses on data that we have gathered in the past, we are on good solid footing in making those numbers reflect reality.

Second, we try to involve people who have a really good idea of these things. These are people who have experience with the cost or profit center, people who are experts on cyber risk, people who have done this before or something similar. With data, they can confirm if it is accurate and relevant. And if data are not available, they will be able to give you good enough estimates of values for costs, benefits, and risks.

Third, combine estimates and guess. One guess from one source is good, but many guesses from different sources is better. Calculating averages is a good way to consolidate or unify estimated values.

Fourth, incorporate the time element. Whenever we guess a future cost or a future benefit, it is important to think about the future environment or scenario that we are envisioning and how that affects the future values of these costs, benefits, and risks.

Last, note that we are not after precision here. We do not need exact numbers to the cent. It is enough to have values (which may be relative) that allow us to differentiate between alternatives or different courses of action. All we really need is good enough information to help us make these good decisions.

## 10.6. Conclusions

Cybersecurity is omnipresent in all aspects of businesses—both as a necessary component in the process of creating and delivering products and services (i.e., as a cost center) and as quality aspects that provide better value and more profit from those same products and services (i.e., as a profit center). But just like any activity in a business, any cost incurred in cybersecurity needs to be rationalized in terms of the benefits it may bring to the business. However, benefits of cybersecurity come in two distinct forms: traditional benefits like profit and benefits in terms of a decrease in risk to a business. Entrepreneurs can use ROI, its modified version in RROI, and NPV to rationalize spending on cybersecurity, both as cost and profit centers. Although there is a lot of estimating involved in their calculations, precision is not necessary for economic evaluation of spending in cybersecurity. It may be sufficient to simply be precise enough to differentiate between alternatives or courses of action.

# 11

# Cyber Insurance for Entrepreneurs

Howard Miller

## 11.1. Introduction

This chapter is about "Cyber Insurance 101 for Entrepreneurs." Before we talk about the details of cyber insurance, let us start with what it is and why you would need insurance to begin with.

Before we can answer that question, we have to put this into the context of managing risk because insurance is part of managing risk. So let us begin with risk. Risk could be defined as "a measure of the negative effect of uncertainty on achieving objectives" [1]. According to the Merriam-Webster Dictionary, an entrepreneur is someone who "organizes, manages, and assumes the risks of a business or enterprise" [2]. If you are going to assume the risk required to succeed as an entrepreneur, would not you want to minimize the risk of failure as much as you can? When you think about it, the key to being able to succeed over time will depend on how successful you are at managing risk in pursuing your business objectives.

How do you manage risk as an entrepreneur? Well, that is the subject of risk management. In the pursuit of reward, you will always have risk, and the specifics of your business environment and your organization will always be changing. So how can you get risk management done? The truth is that risk management is never done and neither is the work of being an entrepreneur until you stop doing business. Now that you realize that risk management is part of the process, the best approach is to make this part of your company's culture and learn about risk management so you can best protect what you are building.

What is risk management? The National Alliance for Insurance Education and Research defines it as the "process of managing uncertainty of exposures that affect an organization's assets and financial statements using five steps" [3]. Those five steps are:

1. Identification and analysis of exposures.
2. Controlling the exposures.
3. Financing of losses with external and internal funds.
4. Implementation.
5. Monitoring of the risk management process [3].

Sounds complicated, but when you break it down, it is understandable.

The most important step in managing risk is identification because if you are not aware of what is exposed to risk, how could you do anything about it? These can be called exposures. An exposure can be defined as "a situation, practice or condition that may lead to an adverse financial consequence; an activity or resource; people and assets" [3].

Working with an insurance counselor as opposed to an insurance salesperson, you may also be able to get consulting and insight into helping you identify your risks. An insurance counselor can match contract requirements and exposures with insurance coverage as well as analyze insurance policy form language. An insurance salesperson sells insurance policies but may not provide consulting. With a product as complex as cyber insurance, it can be very helpful to work with an insurance consultant that specializes in cyber insurance. Consultants, attorneys, CPAs, HR consultants, safety consultants, and security consultants are all examples of professionals that may help you identify risks in your organization. Risk can change over time, so going back to this first step helps find risks that could jeopardize your business that you may not have identified. The most expensive way to address risk is to be surprised and unprepared. This can also be the fastest way to lose your business.

The next step of controlling exposures has to do with what you can do to minimize risk from occurring or, if it does occur, minimizing the damage. Keep in mind that even if you spend money and resources to control risks, you will likely reduce or mitigate it, but not eliminate it completely. There is still a certain amount of risk left over, which is called "residual risk." So what can you do? One of the best tools for entrepreneurs is to transfer the financial cost of risk to an insurance company in exchange for a premium. This is the third step of the risk management process identified above.

Because many entrepreneurs are purchasing insurance for similar risks, the insurance company expects that only a certain number of entrepreneurs will suffer a loss and that overall the group will pay more premiums than the total of all losses for the group. This allows an insurance buyer to pay a small amount of money (the "premium") for a much larger limit of insurance. So that, if you do have a loss, you have transferred a greater amount of financial cost to the insurance company who, less a "deductible or retention," will pay for the damage up to the limit, subject to the terms, conditions, and exclusions of the policy. Having an insurance company cover the financial cost in the event of a loss can mean the difference between staying in business or not for many companies.

A bonus is that insurance companies really want to minimize you having a loss in the first place. Because of this, they may be willing to help you with the control of risk and the resources to respond to risk, which can be extremely valuable in addition to the insurance itself. Always make sure you understand and consider taking advantage of the resources available to you through the insurance company as a policyholder. These resources can benefit a company by reducing the possibility as well as the extent of the damage. Offerings are provided and/or provided at a reduced cost and can apply before, during, and after a loss.

Before a loss, the focus is on reducing the risk. Examples include vulnerability analysis and limited consulting on information security and information security compliance issues. According to Verizon's 2022 Data Breach Investigations Report [4], 25% of total breaches were from social engineering attacks. The "human element" including errors and the misuse of privilege accounted for 82% of analyzed breaches over the past year. To prevent losses, insurance companies may offer proactive security awareness training to executives and staff to avoid and minimize errors that lead to compromise, malware infections, and sending money to unintended third parties. Tabletop exercises can be used to simulate your organization's response to cyberattacks and identify potential weaknesses before you have to actually respond. Online educational portals and technical resources including incident response templates, sample policies, and current news related to understanding and protecting your organization are other useful resources. Other resources include discounts and connections to help you identify, evaluate, and deploy user security such as multifactor authentication, endpoint security detection and response protection, patch management for software vulnerabilities, and ongoing detection of vulnerabilities that could compromise your network.

Of course, you are going to need to implement and monitor the plans and programs to get this done (the fourth and fifth steps of the risk management process). This means that you must know what measures you need to put in place to mitigate risk and what and how much insurance you need to purchase. You must implement these plans and programs to be successful. If you do not complete this last step, you will not minimize the risk or financially protect your business with insurance. Remember that we discussed that your business and the business environment is always changing. To be truly successful, you will need to monitor and reevaluate. This will take you back to the first step of identifying risk, and the cycle will repeat [5]. Consider this process part of protecting something you really care about, your business, its impact on the community, your employees, and your profits.

## 11.2. What Is Cyber Insurance?

Cyber Insurance is an insurance policy designed to respond to the risks associated with confidential or protected information and digital operations. There are different types of information. Examples include, but are not limited to,

- Financial
- Personally identifiable
- Corporate
- Health
- Credit card
- Trade secrets
- Other intellectual property
- Sensitive correspondence and others

As you can see, all of this information is extremely valuable for entrepreneurs' day-to-day operations and to analyze and use for business decisions.

Consider if it is possible to do business when all of your information is publicly available, including your banking, legal, accounting, customer information, and correspondence. You can see it becomes extremely problematic to function when you cannot maintain confidentiality. How could your clients trust you? How could you keep sensitive information away from your competitors? What if all of this information was altered or corrupted so that you could not trust your own most valuable information? How could you do business when you cannot trust your own information to be accurate? What if your information was destroyed or you were denied access to it? How could you function as a business without your information?

These concepts are all embodied in the principle of information security called the CIA Triad, which stands for Confidentiality, Integrity, and Availability. It is clear that if we cannot maintain a certain threshold of these three areas, the structure falls apart and with it your ability to transact business. So you can see how important it is to manage information security risk. Using cyber insurance is part of your toolkit for managing cyber risk and protecting the success of your business.

After reading the above, I hope you start to consider and understand how valuable confidential or protected information actually is. Like other assets in your business, they can have significant value to you, but they can also pose a liability when they actually or allegedly cause harm to others. When it comes to liability, remember even if you do not believe you are negligent in harming someone else, you may still need to defend an allegation of negligence. This is one of the important reasons to have liability insurance.

## 11.3. When and How Do I Buy Cyber Insurance?

### 11.3.1. When Should I Buy Cyber Insurance?

You should buy cyber insurance when you want to fully manage your cyber risk. Part of fully managing your cyber risk includes controlling cyber risk and also financing cyber risk, which for most entrepreneurs is transferring the risk through insurance.

There are a number of answers as to what is the right time to buy cyber insurance. Insurance is for companies that have something to lose. So your operations or assets are valuable enough that you want to insure them. If you have something to lose, then the cost of a loss could be more than your company could bear. It could be a lot more bearable if you were able to transfer part or most of the financial cost of the loss to an insurance company in exchange for a premium. When you are in a growth mode, you do not want a loss to affect you financially to the point where it could destroy or impede your goals. This is a good time to take advantage of insurance. Another reason is when you have insurance in place, your insurance policy is coming up for renewal and you want to continue it for another year. These are all good reasons to purchase cyber insurance.

## 11.3.2. How Do I Get Cyber Insurance and Who Should I Contact?

Getting insurance is similar to getting your legal, accounting, and other essential business services. Although digital offerings are available, many companies prefer the consultative advice of working with a professional that understands the needs of their unique company and works to provide the best solution.

Since the insurance industry is regulated, insurance companies that provide the insurance are represented by insurance agents that are licensed to sell various types of insurance. These agents can represent one company or can be brokers that can represent multiple companies in order to find the best pricing and terms for your insurance policy or insurance program.

An independent agent can provide multiple quotations from different companies, depending on your risk profile. A bindable quote is an insurance offering that, subject to any additional information or verification, is available to purchase.

## 11.3.3. How Do I Apply for Cyber Insurance?

Applying for insurance will require an application that includes questions as to the nature of the risk and the extent to which that risk is minimized by appropriate controls. Keep in mind that the questions on the application are specifically designed to determine if you are implementing proper cyber risk controls. As such the application can be helpful as a checklist when it comes to self-assessment.

The insurance companies deal with speculative risk, meaning they could make a profit or lose money depending on how many insured have claims and how severe the losses are. Because of this the underwriting is designed to determine your company's potential for losses and charge accordingly. In some cases, an insurance company may not insure you at all because they believe insuring you will be unprofitable based on their experience or analysis.

Let us look at an example and related risk controls. Let us say you are a manufacturing company with $25 million in revenue. Your ability to protect your organization, meaning your ability to control the risk, is average. In our example a small business insured buys a $1 million limit cyber liability insurance policy for, say $20,000, and suffers a ransomware loss. Let us assume the following conditions: average defense to protect the organization, average ability to detect a cyberattack and recover from it, attackers are of average sophistication and motivation, and the information subject to the attack is slightly below average. The estimated loss was $1.4 million dollars in damages including about $300,000 due to the disruption of operations; about $400,000 in financial loss, fraud, claims fines, and reporting; and the rest due to information discovery, mitigation, and liability. This example was taken from a cost calculator [6], and there are several to choose from, but the results may vary compared to an actual attack on your organization. Using this example, you could pay a $20,000 premium for your policy and the insurance company would pay a $1-million-dollar policy limit loss. You can see why risk controls are so important in minimizing claims and determining pricing by an insurance company.

Basic factors in pricing and applying for cyber insurance include the amount, location, and sensitivity of the information you have, what industry you are in, your company's gross revenue, and whether you are in compliance with industry and regulatory frameworks that apply to you.

Advanced factors include speaking with underwriting to better understand your security posture and insurability. Some companies may want to benchmark their current limits of insurance with those of their peers to determine how much insurance similar companies are purchasing. Quantifying what your potential loss and probability might be and weighing that against the cost of the insurance and how much of the damage your company should absorb could involve analytic and financial modeling.

An insurance consultant can advise on the differences in policy forms. Keep in mind that insurance policies are legal contracts, and specialized attorneys that focus on insurance contracts called "coverage counsel" can be used to analyze and provide legal expertise regarding insurance contracts.

Once you select the right policy, limits, and deductibles/retentions, then you can make a premium payment and bind the policy.

# 11.4. What Are Some Controls That Would Be Important?

## 11.4.1. Network Security Vulnerabilities

Checking for assets with insecure configurations, outdated software, or other vulnerabilities, and fixing these is important. These could be determined by an external scan that could identify weaknesses of outward facing security architecture. Keep in mind that most insurance companies will perform this external scan as part of their underwriting process. They may share the report with you. If the results are unsatisfactory, you may not qualify for a policy.

## 11.4.2. Email Security

Most companies become victims through email. They could be tricked into releasing money, downloading malware, or the loss or disclosure of confidential information. Do you tag your emails coming from external sources? Do you use a service to screen your emails for malicious code? Do you employ technologies that determine if the email is genuine and it is coming from who it claims to be coming from, such as SPF, DKIM, and DMARC?

Because so much information is contained within an email, it is not enough just to have a username and password. Effective security is "defense-in-depth" so that, if one layer fails, there is another layer to prevent compromise. This is why multifactor authentication (or MFA) is so important not just for email but as a second layer of defense for information access in general.

## 11.4.3. Internal Security Controls

Putting a lock on your sensitive data or encryption improves confidentiality and puts you in a more defensible legal position if your data are compromised wherever it is located. Protecting your endpoints with advanced tools such as next-generation antivirus and endpoint detection and response are highly recommended. These tools not only identify malicious code or attacks but can also respond to mitigate damage.

Other considerations include the following:

Privileged account access management software to control high-value account access to your information. Two concepts are important. The first is "least privilege." This means that a user should only get the minimum amount of access necessary for them to complete work [7]. This way you limit access to more information than necessary and reduce risk. The other principle is "need to know," where only information they need is granted. Need to know refers to the relevance of the data involved in the role they play or the job function [7].

User behavior analytics is a way to monitor unusual behavior across your network which could be an indicator of compromise compared to what is normal for your organization based on factors such as time of day, location, type, and amount of data.

Patching software vulnerabilities in a timely manner before loss is another important practice. Criminals will continue to scan for unpatched vulnerabilities and exploit them until they are no longer available. In January 2021, the Microsoft Exchange server was exploited due to a vulnerability that affected over 30,000 United States (US) businesses. Microsoft provided a patch, but the damage was already done for many companies. As massive attacks like these become public, insurance companies will likely ask if your company had an exposure to this loss and if you patched your systems and how fast did you patch it [8].

As software nears its end of life, patches may not be available or will disrupt the function of the software performance. In this situation it is important to discontinue this software or at least segment it from the rest of your information environment. This same concept can hold true for IoT devices and Operational Technology (OT) where the hardware utilizes software or firmware that is difficult to update or was not designed to connect to information technology systems with the same resiliency. This cyber-physical threat can take down or compromise a device, manufacturing facility, or even the critical infrastructure of our nation's water, power, or other systems.

Finally, do you utilize SEIM, SOC, and vulnerability management tools? If so, how often are they monitored and by whom?

## 11.4.4. Backup and Recovery

Backup and recovery are critical if you want to maintain continuity of operations. Otherwise, you can lose your data either due to corruption, deletion, modification, encryption, or destruction, and your operations will be disrupted. The length of time of that disruption will depend on your ability to recover your data from backups including the programs, operating systems, and configurations needed to resume operations in the manner before the loss occurred. How long would it take you to recover from your backups? How much time between the time of loss and the date and time of the last backup can you afford to lose? If you lost an hour's worth of data, could you recover? What about a day, week, or month? You can see how important backups are. The questions about maintaining the integrity, availability, and confidentiality of your backup are of concern for an insurance company that could have to pay for the cost of recreating your data, the lost profit and extra expenses associated with a digital operation disruption, and the loss of confidentiality of your stored sensitive data.

Are your backups segmented from an Internet connection so they are not affected by malicious code? Are they stored in the cloud? Are they encrypted so they are not readable in plain text? How are your cloud backups protected? Do they have separate credentials? Do you use multifactor authentication as an added layer of defense? How old are they and how often do you verify and validate the recovery of your backups?

## 11.4.5. Phishing

We live in a world of interconnected systems that include software, data, hardware, and human elements. Failure to address the human element of security can lead to compromise even if you have strengthened other aspects of your organization's security program. This can lead to a compromise of digital security or monetary loss. The key component in mitigating this risk is to increase awareness and education at all staffing and executive levels to better recognize and minimize errors. This social engineering training keeps people from falling victim to scams and deception that can allow a criminal to steal money and compromise the system.

When it comes to money, the key is a process that includes documentation for all wire transfers with written authorization. A protocol should be in place for verifying any change requests or funds transfer instructions and requests using a different method than the original request, meaning if you receive a request to change wire instructions from a client by email and that you use the original contact information to call the client and verify the change. If the email is altered or initiated by a cybercriminal posing as your client, then responding to the same email is simply communicating with the criminal. Calling the customer can reveal that the request was fraudulent, and a loss can be averted.

**FIGURE 11.1**   Peter gets locked out of his computer.

Illustrated by Phillip Wandyez.

*Peter had been busy on the Internet again, buying things from shady websites. One day when he turned on his computer, a message popped up telling him that his computer was locked and that the only way to access his files was if he paid a ransom in cryptocurrency.*
*"Cryptocurrency? I barely have any real currency!" Peter exclaimed. "What is this, some sort of ransom?" That is exactly what it was. Peter's computer was infected with ransomware, locking him out of his computer and causing his business operations to grind to a halt. Of course, had Peter purchased a cyber insurance policy, it is possible that the losses in revenue and other costs associated with the ransomware incident could have been covered.*

## 11.5. What Are Some Important Contractual Aspects to Know about the Insurance Policy?

The cyber insurance policy is a written contract between your company and the insurance carrier that states the agreement, rights, and duties of both parties. Keep in mind that insurance policies have obligations. Failure to follow the obligations could potentially get your claim denied. Insurance policies can also have requirements for protective safeguards and the insurance application becomes part of the insurance contract. If statements in the application were false, then the insurance company could state that they had relied on those statements to underwrite the risk. A material misrepresentation could be grounds to deny coverage. Word of advice: Be careful about the claims you make as to your controls and procedures. If you say you update your software every week and then have an insurance claim where it was clear that you failed to do so, it could be considered a material misrepresentation.

## 11.6. What Are Some Important Parts of the Insurance Policy to Pay Attention To?

Some important parts of a cyber insurance policy include conditions, definitions, exclusions, and insuring agreements.

Conditions can include when the policy will respond to a claim that might not have been discovered yet and comes to light in the future. It will also explain the defense and settlement provisions such as the right and duty of the insurance company to defend you legally, what happens when the insurance company can settle the claim but if you refuse to settle, your authority to agree to a settlement without the insurance companies' approval, and what happens if part of the claim is covered and part is not.

The policy territory for a cyber liability policy should be worldwide as the Internet is worldwide.

The notice and claim reporting provisions outline your duties under the policy to inform the insurance company and make a claim. This can include both actual demands for damages but also potential claims that have not fully materialized yet. Make sure you preserve your rights under the policy by paying attention to these obligations.

Other aspects include how the policy responds if you acquire new companies, merge, or sell your company, how the limits and deductibles apply, and how other insurance policies may apply. You may see over a dozen limits and insuring agreements on cyber insurance policies. Keep in mind that all of this coverage is usually limited to a total amount called an aggregate limit. Understanding your aggregate limits makes it clear how much total insurance coverage you have. If you have a one-million-dollar aggregate limit and you have four insuring agreements each with a million-dollar occurrence limit, you still only have a total of one million in the aggregate for all four insuring agreements.

Two important points regarding definitions and exclusions: First, how words are defined in the policy could actually take away coverage. Consider what the initiation of a computer system and other words mean. If they are narrowly defined it can minimize your coverage. Second, exclusions can carve back coverage by excluding coverage and then stating a word like "however" which narrows how the exclusion applies and can give back coverage.

The insuring agreements state what you are covered for. On a cyber liability insurance policy your insuring agreements are covering wrongful acts you could be actually or allegedly liable for and are triggered by causes of loss, or what are called "perils" in the insurance industry.

Take a look at the definitions of some cyber liability insurance wrongful acts and consider if the perils that trigger these wrongful acts are relevant to your business or if anything is missing. By doing this you are now starting to manage risk by identifying the cyber-related causes of loss that could impact your business. Identification is the first and most important step in the risk management process.

Coverage is more affirmative when it clearly addresses the exposure and cause of loss and is not taken away by any definition or exclusion. This allows for increased confidence by the insurance buyer that they will be protected. Keep in mind that insurance companies offering other policies such as property, general liability, and crime have been involved in litigation to determine if these other policies will pay or cover a cyber-related loss. This type of litigation will continue in the future. A cyberphysical risk may cross both physical and digital boundaries, parts, and systems. Policy forms will change, and insurance companies will respond by adding policy endorsements excluding, broadening, or clarifying how a given policy should respond to cyber-related risk. The sophistication of an attacker, threat surface, and the use of computer automation and artificial intelligence will continue to evolve. It is recommended to err on the side of clarity when it comes to any insurance policy that could involve a cyber-related claim.

# 11.7. First- and Third-Party Insuring Agreements

Cyber insurance has two key areas. One is called "first party" and has to do with your company's direct and indirect costs associated with a cyber incident such as a breach of confidentiality or a cyberattack that disrupts your operations. The second area is called "third-party" liability. This is your liability to a third party, primarily for wrongful acts associated with the use, disclosure, access, or theft of confidential or protected information. Cyber insurance has expanded to include monetary loss associated with your money as well as media liability exposures.

## 11.7.1. First Party

First-party insuring agreements pay for your company's costs associated with a cyber incident. Cyber risk can impact an organization in a number of ways including disruption of operations, damage to reputation, and financial costs associated with assessing and mitigating a cyberattack or breach of confidential or protected information.

*Breach Costs and Crises Management.* According to the Verizon Data Breach Investigation Report, about 28% of breaches were discovered months after compromise. In 2021 about 49% were discovered in days, not months. Keep in mind that more companies are being notified by criminals looking to collect a ransom before the release, destruction, or theft of a company's valuable information. This is not the same as the organization being able to detect the compromise itself, which is much more proactive [4].

According to the IBM/Ponemon report, the average time to identify and contain a breach is 277 days. The biggest cost saver was realized by the companies that used an incident response team and tested their programs [9]. Crisis management or breach response coverage can provide and pay for a computer forensic investigation. This can be expensive and the importance of knowing what was affected, accessed, or exfiltrated can be important in what comes next. You may need to notify affected third parties as required by law or voluntarily. Once affected parties are notified, you may need a call center to respond to the inquiries. Credit card monitoring can be provided to affected individuals whose personally identifiable information was exposed or stolen. Legal advice is extremely valuable during a crisis, especially as cyber risk can touch on a patchwork of local, state, federal, and foreign regulations and laws that may require certain responses in specific time frames to be compliant. To mitigate damage to your reputation, PR firms are provided to improve the messaging to stakeholders.

For many companies, it is not just that cyber insurance pays for these costs, it is the ability to access these resources at a time of crisis provided by the insurance carrier that makes all the difference in being able to survive financially and reputationally and minimize the impact of a cyber loss. In some cases, the resources available through the policy create a robust response and support your incident response plan. In the same way that the signature on a life insurance policy can create an instant estate, the binding of a cyber insurance policy can create a suite of experts that help you respond and recover from a cyberattack.

*Business Interruption.* A business interruption has to do with the lost profit and extra expenses arising out of an operational disruption. It is typically triggered due to direct physical loss or damage to tangible property. Business interruption coverage is sometimes called "time element" because the amount of damage is correlated to the amount of time a business is shut down or disrupted.

The same coverage can apply to cyber insurance, except here we are dealing with data affected by a cyberattack rather than tangible property such as buildings, inventory, and equipment. To illustrate this, we could take a company providing professional services such as a law firm or accounting firm affected by ransomware. The ransomware encrypts their data and operations come to a halt as the company is unable to use their information systems and data. It soon becomes clear that it is not just the value of recovering the data itself, but every hour that goes by is lost revenue, damage to reputation, and potential legal action by affected third parties. Business interruption coverage can pay for the lost profit and extra expenses to get back up and running due to interruption to a company's systems and can extend to other systems (called contingent or dependent systems) due to a covered cause of loss. Now ransomware typically encrypts or locks your data to extort a monetary payment from you. This does not mean that information has not been exfiltrated, which could be used to extort money from a company in the hopes that it will not be released, or it could be sold without a company's knowledge. Unauthorized access to confidential or protected information of others in your care, custody, or control can open a company up to liability which is addressed in the third-party liability insuring agreements.

*Cyber Extortion.* Cyber extortion is payment to prevent digital destruction, impairment, or disclosure. Paying money to a criminal may not be legal if it is attributable to a country under sanction by The Office of Foreign Assets Control (OFAC) of the US Department of the Treasury. That being said, many companies pay a ransom because they believe it is the best course of action to minimize financial damage or their only hope in recovering their data.

Attacks by nation states can be carried out through kinetic attacks as well as cyberattacks. As the intensity of war affects the geopolitical climate, insurance companies can be concerned with a loss that could affect a large number of insureds at one time. This aggregated exposure could be financially catastrophic for the insurance industry. Policy language regarding war exclusions and coverage for cyber terrorism is in flux and something to pay attention to in any cyber insurance policy. Refer to the import parts of a policy to pay attention to in this chapter. Also consider how words are formally defined in the insurance contract or if there is no clear definition. Some policies are "silent" on whether coverage is specifically covered or excluded, which may or may not be favorable depending on the loss scenario. In general, clarity is good, but cyber risks can evolve rapidly, matters are litigated and tested in the courts, and new risks emerge. Policy language will continue to evolve in this environment.

## 11.7.2. **Third Party**

Confidential or protected information from its inception to disposal is an asset, but it can also be a liability. As entrepreneurs do business, they will both create and acquire information from third parties throughout their business life. Even if that information is stored by a third party such as in the cloud or handled by a company that may process it, damage or loss of control of that information can have a serious effect. Businesses have a duty to protect their clients and employee's information.

Legal exposure can come in the form of local, state, federal, and international laws, regulations, nondisclosure agreements, and contractual requirements with clients and suppliers. Keep in mind if a subcontractor is compromised while storing or processing data, such as your client's data, you can still be responsible for the loss.

Third-party liability can pay for legal defense and settlement due to a claim brought against your company for actual or alleged damage. Different liability insuring agreements are triggered by different causes of loss.

*Network Security and Privacy.* This is liability arising from the transmission of malicious code, such as ransomware, denial or failure to prevent unauthorized access to computer systems, or the use of a computer system to carry out an attack against a third party or identity theft.

Privacy triggers may be related to a violation of a data protection law, your privacy policy, or a nondisclosure agreement. Some ways this can occur include loss, theft, unauthorized use, access, disclosure, and failures in maintaining control of information. Privacy law varies across countries, states, and local jurisdictions.

The definition of what is considered personally identifiable, and reasonable security requirements needed to protect private information change over time. Human rights and freedom can be associated with a person's right to privacy. The definition, protections, and balance of these rights with national security are an evolving landscape, which may influence liability in different geographic locations.

Network security and privacy liability coverage can be valuable in providing specialized legal expertise in defending these claims.

*Media Liability.* Many companies engage and provide content over the Internet. In the course of creating, disseminating, or releasing content to the public, defamation claims and infringement on intellectual property such as copyrights and trademarks (but excluding patents) can lead to media liability claims. Those that are in the media industries such as publishers, creators, and producers of content may have specialized needs that could go beyond the standard media coverage offered on a cyber insurance policy.

*Technology Errors and Omissions.* If you are a technology company, you may provide a technology product or service. Errors and omissions are for professionals that have a higher degree of knowledge and expertise than the general public. Failure of a product to perform or negligence in providing or failing to provide a technology service can lead to lawsuits by third parties seeking financial damages. This covers a range of industries such as technology manufacturers, distributors, software developers, cloud service providers, and information security professionals, as examples.

*Regulatory Defense and Penalties.* Regulatory investigations, fines, and penalties can be brought by regulators tasked with enforcing these laws or the safeguarding of computer systems. Governmental authorities have the right to investigate and can bring fines and penalties against your organization. This coverage can provide legal defense against actions brought by regulatory authorities and the cost of associated fines or penalties. Examples of relevant regulations include the General Data Protection Regulation (GDPR) related to data of European Union citizens. Others include the California Consumer Privacy Act (CCPA) related to information collected on California consumers, and the Health Insurance Portability and Accountability Act of 1996 (HIPAA), protecting individual's patient data. Industry standards such as Payment Card Industry Data Security Standard (PCI DSS) cover the companies that accept, process, store, or transmit credit card information. Coverage related to the defense, investigation, assessment fines, and penalties can be related to this coverage or listed under separate insuring agreements. If your company is subject to regulation based on your industry or the type of information in your care, custody, or control, check your policy to verify what and for how much you are covered.

## 11.8. Conclusions

Think of where insurance fits into the overall risk management process. By using cyber insurance as a tool in managing cyber risk, you can better protect your company financially, leverage the resources provided by the insurance company to minimize the cost of a cyber risk event, and become more resilient. The advice in this chapter was designed to give you a better understanding of how you are exposed to cyber risk, what cyber insurance is, when to get it, how to apply and get the best value from your cyber insurance policy, important aspects of the cyber insurance policy contract, and insuring agreements. This knowledge can help you grow and sustain your business by utilizing cyber insurance as part of your risk management program.

# References

1. OCEG, "Glossary," Open Compliance and Ethics Group, 2022, Retrieved November 26, 2022, https://www.oceg.org/glossary/.

2. Merriam-Webster, "Entrepreneur Definition & Meaning," n.d., Retrieved September 11, 2022, https://www.merriam-webster.com/dictionary/entrepreneur.

3. National Alliance, "Risk Management Essentials," 2nd Edition, Digital Version, The National Alliance Research Academy Risk and Insurance Studies, Austin, TX, 2014, International Standard Book Number: 978-1-878204-77-6, https://nationalalliancebooks.com/Studies.

4. Verizon, "Ransomware Threat Rises: Verizon 2022 Data Breach Investigations Report," 2022, Retrieved 11 September 2022, https://www.verizon.com/about/news/ransomware-threat-rises-verizon-2022-data-breach-investigations-report.

5. Miller, H. and Griffy-Brown, C., "Evaluating Risk for Top-Line Growth and Bottom-Line Protection: Enterprise Risk Management Optimization (ERMO)," *Environ Syst Decis* 41 (2021): 468-484, doi:https://doi.org/10.1007/s10669-021-09819-x.

6. Hiscox, "Cyber Exposure Calculator," 2022, Retrieved November 28, 2022, https://www.hiscoxgroup.com/cyberexposurecalculator/.

7. Chapple, M., Stewart, J.M., and Gibson, D., "(ISC)2 CISSP Certified Information Systems Security Professional Official Study Guide," 8th ed., Sybex, 2018, 105, 283.

8. UpGuard, "Biggest Data Breaches in US History [Updated 2022]," Upguard.com, 2022, Retrieved 11 September 2022, https://www.upguard.com/blog/biggest-data-breaches-us.

9. Ponemon, "Ponemon Report Cost of a Data Breach 2022," IBMM 2022, Retrieved October 4, 2022, https://www.ibm.com/reports/data-breach.

# Disclaimer

# 12

# Cyber Resilience for Entrepreneurs

Paul E. Roege

## 12.1. Protection versus Performance

Cybersecurity experts tell us to use special tools and practices in order to keep our systems pristine; to many of us, the rules seem mysterious and endless! They create more work, making it more cumbersome to do business, and it is easy to lose track and get behind on them. Worse yet, there is no perfect protection. Arguably, the most effective way to protect yourself from a cyberattack would be just to turn off your computer—which may sound tempting, but probably would not increase sales or product delivery. Is there a practical way to deal with the bottomless pit of cyber risks that also allows you to focus on growing your business?

Many people think of their business as a child. You conceived it, birthed it, and put your heart into growing the business into a strong, productive entity. Like parenthood, it would be easier just not to take on the challenge—but for an entrepreneur, missing the game is not an option. The parallel between parenthood and running a business is not limited to the challenge and deep commitment. We need parents for society to thrive, just as entrepreneurship is the engine for a nation's economic survival and growth. Let us explore how our understanding of parenthood can help to protect and grow your company, even in the face of cyber—or, for that matter, other—risks.

Even though we all were once children and have observed other parents—maybe even found time to read a book or two—most new parents feel at a loss at some point when it comes to dealing with some situation. How do you protect your child from being hurt while giving them the chance to learn? How much supervision do they need? When should you correct them? And, of course, should you take them to the doctor this time? There should be—and are—checklists and guides to help navigate those daily situations, but you can never follow them all, and they still do not cover every possibility. A critical lesson for parenting success is that of humility. Once we admit that we cannot know everything, nor predict or control the future, we open a dramatic opportunity for success—or, alternatively, anxiety... It is a choice.

Once we recognize our limited ability to control, what determines our future pathway? Is there a strategy that at least increases the chances of positive outcomes? The answer is as old as life itself: resilience. The term has become popular in recent years, although few understand it. Some talk about energy or climate resilience; others use the term to address infrastructure systems or socioemotional well-being. What do all of these things have in common with the evolution of primordial proteins and cells that formed, reformed, and reproduced in ancient oceans to eventually produce today's diverse, complex world? The answer boils down to a basic process.

## 12.2. Introducing Resilience

Resilience describes the capacity of a system to deal constructively with change. Accepting our limitations, resilience depends upon our ability to anticipate and prepare for change, absorb, and respond to its effects; restore basic capabilities; learn from the experience; and use that learning to adapt for greater future success. Charles Darwin recognized this pattern when he observed that biological species evolve to survive and flourish in their constantly changing environment. Through naturally varying processes, organisms are created with slightly different features; those that are better suited to the local environment thrive and reproduce. Taking that a step further, beings with a greater capacity to adapt are that much more successful as things change. This natural adaptation process does not require organisms to be intelligent. Drawing upon that example, we humans have developed the experimental process to increase understanding of unknowns. Clearly, this is a "no-brainer" idea, considering its proven success (over millennia). We, however, do not want to take a "no-brainer" approach because we do not want to take millions of years—or suffer mass extinctions—to arrive at sustainable solutions. How can we use this resilience-building process efficiently to help our businesses weather looming cyber threats?

Let us start by thinking through the experimental process. From the time we are born, we observe things and our brains notice patterns. We hypothesize about how things work—for example, "when I cry, I get fed." Feedback serves to reinforce, rebuke, or modify our understanding. This is an ongoing process because, inevitably, things change. We learn to look for more nuanced signals to improve predictions. "Is mom smiling? Do we have company? Are people putting on their coats?" Such signals can help us to anticipate different responses, changes, or events. As we grow, we learn new ways to cope with hunger—perhaps exploring the refrigerator or, ideally, learning to cook.

The challenge of parenting (and of entrepreneurship) is that we are no longer able to rely upon someone else to take care of us; we experience role reversal. It is quite a different situation to be hungry versus hearing one's new baby cry. Not only may it be hard to figure out why the baby is crying; the sound itself can be stressful, which can degrade our perception, judgment, or patience. It is time to take our resilience skills to a new level. As parents, we find ourselves balancing finances, relationships, health, and so many other dynamics, taking responsibility for the outcome. It is worth exploring how proactive resilience (learning) can be more effective, yet simpler, than reactive protection (following the rules).

If parenting can illustrate entrepreneurial challenges, looking at health and medicine may be useful to contrast resilience and protection. Over the past century, especially, medicine has become largely reactive, particularly in the "developed world." We rely upon others (government and trade organizations) to control the quality and safety of food, water, and our environment. With this huge, bureaucratic network of rules, people, and checks, the rest of us do not "need" to worry unless something goes wrong. When something becomes noticeable—like "my head hurts"—we have a headache. Take aspirin, Tylenol™, or, now, Advil™. If we have some other problem, we treat the symptom (bleeding, swelling, etc.), possibly follow up to identify the direct cause, and then take the right medicine or go in for some curative procedure. Even preventative measures, such as influenza vaccines, target specific risks rather than providing systemic resistance to a range of hazards. Scientific advances have given us a sense of security about our knowledge and the robust nature of our targeted solutions.

Three problems with this arrogant approach to medicine are that our bodies are complex, the threats are even more complex, and the world is in a constant state of change. The reactive approach to treating symptoms and illnesses bypasses opportunities to identify and address situations before they become worse, and it creates a "whack-a-mole" situation in which we spend a lot of money, effort, and emotional distress trying to patch a deteriorating system. Even more scary is the fact that viruses and bacteria are living organisms that, even with their "no-brainer" approach, are resilient. When scientists produce vaccines that target a particular pathogen, the "bugs," which have evolved via natural selection to be adaptive, do just that. The growing threat of drug-resistant pathogens has been catalyzed by reactive medicine displacing the health approach.

**FIGURE 12.1**   Peter's social media gets hacked.

Illustrated by Phillip Wandyez.

*In an effort to appear "hip," Peter tries to advertise on all of the latest social media platforms. One day, he gets a call from one of his friends who follows Peter's page. "Peter, you are posting some really weird stuff on your page," he said. "What do you mean?" Peter asked. "I haven't posted anything since yesterday." Peter checked his page and someone had gained access to it and was posting under his name. "Oh great, just what I need! What should I do?" Peter wondered. Peter had never thought about what would happen under a scenario like this and had no plan for how to respond.*

## 12.3. Holistic Approach

Fortunately, there has been a recently growing trend back toward *health* versus *medicine*. When we are healthy, our bodies become stronger and less vulnerable to malaise, and we recover more quickly from injury. Healthy living, which includes measures like monitoring key indicators, managing our nutrition, and balancing exercise and rest, has proven much more effective toward survival and well-being, compared to reliance on reactive medicine.

But you are not reading this book to become healthier or a better parent, we are here to figure out how to run a business in the jungle of cyber threats. Can we relate the past few paragraphs so that you did not waste your time? So far we have compared building a business to raising a child and advocated for healthy living versus taking medications. Before we move on, let us explore one remaining important idea: the focus on outcomes. Back to parenting, the goal is not necessarily to have Band-Aids™ in the drawer nor even to make sure your child never falls. What we really seek is to live a long, satisfying life—for our children to go even farther in life than we did. Similarly, our ultimate company goals are not to have nice computers, or even to keep the power on, what matters is that your company grows and achieves its business objectives. Having talked about resilience principles and processes, let us now examine how resilience relates to growing a business in the face of cyber threats.

The first important point is that resilience thinking does not replace discipline. If you build in a flood zone, consume a high-fat diet, or walk on the edge of a cliff, you are ignoring known risks. Resilience does not replace the application of experience-informed standards and practices; instead, it seeks to inform them. In the cybersecurity context, the systems of passwords, firewalls, and patches we use are the result of experience and learning. These rules are there to save us the pain and effort of learning from someone else's suffering. In that light, a resilience approach starts with leveraging the lessons and solutions that have been developed by others. It is important to remember, as we have just observed, that we need to do those things with the outcome in mind. Change your password to reduce the chances of compromise not just follow the prompt. Do not just avoid using your birthdate because that is the recommendation; think about how someone might guess your pattern if they have access to information about you.

Second, think about how your company works, and what "health" would mean. Typically, you deal with people, money, information, production, products, and services. Today, most businesses use computer (cyber) systems and applications to manage each of these—which is why you are reading this book. Cybersecurity programs seek to protect those computer systems from compromise and degradation. But your company's business processes are the foundation for performance and capacity. In order to effectively manage and grow the business, you must have a good handle on how these systems interact, what are important performance metrics, and, conversely, what would be indicators of concern? This is the crux of health as it relates to business. The modern nuance is that, just as computing power can streamline our business processes and accelerate production, computing malaise can turn things south faster than any manual process.

Taking the resilience or "health" approach, we think about our business processes as physicians consider biological systems—for example, pulmonary, circulatory, nervous, and digestive systems. Our brains maintain corresponding processes within certain bounds. That is why the first thing they do when we visit the doctor or dentist is to monitor "vitals." Nowadays, there are millions of fitness watches out there monitoring heartbeat, breathing, oxygen levels, and activities to help us monitor our physical condition and alert us of problems. Especially since most of these devices are online, there are some concerns about cybersecurity related to these devices—although, since the watches do not directly control our bodies, the main issues relate to privacy and security. There are also more sophisticated monitoring and life support systems like pacemakers and hospital ventilators—any of which could become vulnerable to cyber threats. In any of these cases, it can be extremely valuable to find independent ways to monitor important parameters and send alerts, not only to detect potential disruption due to cyber-attack, but to help us notice problems due to any other cause.

## 12.4. Resilience as a Cycle

Keeping in mind this focus on business processes, not strictly on cyber systems, let us review the basic resilience cycle in the context of your company. We want to be prepared to move quickly and effectively through each phase associated with an event or changing condition:

- *Anticipate:* This means being ready for things that you do not expect. Proactively think about different types of change and events and how those might impact your business. This does not mean just looking at "cyber" vulnerabilities but at any processes that are important to your business. Cyber events can impact any aspect of your business as well as resources your company depends upon; moreover, you do not just want to be cyber secure, you want to be resilient. Look for ways to limit negative impacts, especially catastrophic failures, by limiting critical dependencies, identifying indicators that could warn of impending upsets, and planning for responses that can limit consequences or help get things back online. Set alerts to warn of unusual events (transactions, communication disruptions, or variations in production processes), especially involving those core business systems.

- *Absorb and Respond:* You do not want to be shut down or lose your business when some disaster hits. A better way to think about it would be "how can I keep things from becoming a disaster?" Look for key dependencies that could shut you down or severely degrade your operation. What if your email, ordering system, or bill payment system were disrupted? Do you have communications and accounting information backed up? Could you call, visit, or send a check? If the office had to close, could you and your employees work from home or another location?

- *Recover:* This is the ability to get back to normal as quickly as possible. Think about measures you might take to ensure the ability to get back into operation. Not just backup files, but perhaps offsite resources for information, financial, and production processes. Specialized, long-lead machines and software may be a disadvantage here, as well as unique or distant sources. The more ways you have to operate the faster you may be back in business.

- *Learn and Adapt:* Why not take the opportunity to end up stronger than you were before the challenge hit? This is not just a matter of figuring out what went wrong. How can you structure your operational and information systems to facilitate learning and adaptation? Think twice about buying into a proprietary stand-alone system. Could you easily adopt emerging hardware/software technologies? Your business offering and production technologies may change over time—information technologies evolve particularly quickly.

Figure 12.2 illustrates how a resilience strategy not only reduces lost opportunities associated with disruption but can help us end up in a better position if we adapt things based upon our new knowledge. Coming up with a "killer" product or solution is a great starting point, but we need to do more than just protect it, we must strive to improve it. This applies not only to major disruptions but to everyday events and new information. The world changes every day and conditions evolve; you want to be the business that thrives when others may not.

**FIGURE 12.2**   The resilience advantage.

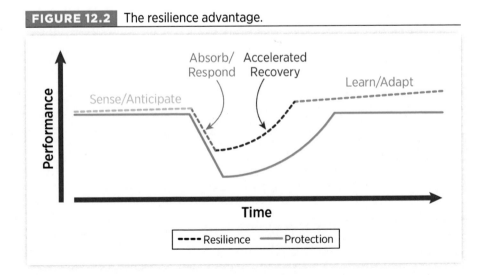

## 12.5. **Design Principles**

Beyond learning how to manage in resilient ways, there are particular design principles that can help make your company inherently more resilient. These apply to any type of system, whether it is, for example, an organization, machine, or community.

*Sensing and information.* Having reliable information available and usable is important through all stages of the resilience cycle, making it easier to limit losses and accelerate recovery. Timeliness, accessibility, accuracy, and validity are, of course, important—a challenge is to balance quantity with quality and value. As the expert, you must decide what information is worth collecting and maintaining to make you more responsive, not overburdened. What customer or market data could best help you identify expectation gaps, competition, or value enhancement opportunities? Are there alternatives or substitutes for key communication or supply chain items?

*Resistance.* This refers to how well your systems are able to weather disruptions. Cyber protection measures and fault-tolerant systems are examples of design features that can help you continue to operate, or limit degradation, in the face of adverse conditions.

*Margins/buffers.* This is a perennial issue in the business world. Buffers allow different parts of a process to function without disrupting each other. Well-honed twentieth-century supply chain managers sought to reduce costs through "just-in-time" concepts in which inventory levels were kept at a minimum. Twenty-first-century managers are rethinking the value propositions as COVID-related restrictions resulted in substantial disruptions to materials, labor, and transportation. The principle applies to any type of process though—for example, energy or water storage.

*Stability.* We design systems to function efficiently under expected conditions, and modern computing systems have enabled a high degree of optimization. Unfortunately, conditions are rarely at the design point, so it is particularly important to understand what happens when things change. In particular, if there are situations when things would change quickly or dramatically—for example, material transition temperatures (e.g., melting) or message volumes—it is important to avoid baking them into the system, or at least have schemes to avoid them.

*Dependencies.* Obviously, anything that prevents something else from happening represents a vulnerability. If you need a mobile phone to validate a transaction, you have a potential single-point failure. If two alternative production options depend upon the same part, the process could be stalled when the supplier has an availability problem. If you cannot avoid dependencies, look for ways to mitigate them, for example, through redundancy (e.g., backup power, equipment, or data storage) or even substitute processes.

*Diversity.* We celebrate diversity when it comes to ideas and culture; the same principles actually apply to engineering and business processes. If you have multiple sources of energy, such as the electrical grid and local generation, you may have more options in case of a power outage. Different kinds of machines may be able to tolerate different conditions. And, yes, diverse people bring different insights when it comes to figuring your way out of a problem.

*Flexibility.* The more you can adjust the less likely you are to be impacted. Some machines do not operate well at different speeds. Could you relocate or work from home if the office were temporarily unusable? Could you produce different products as the market fluctuates or changes?

*Adaptability.* This may seem similar to flexibility but relates more to the ease of changing the system in response to conditions or learning, rather than accommodation of variations. Designing systems that use smaller, modular devices (generators, storage, processors) may be easier to adapt when compared to large, stand-alone systems. "Open" architectures, "cloud" computing, and multiskilled workforces can make it much easier to adjust business processes and offerings as things change.

These ideas of resilience cover a range of facets, from how to think to considerations for tweaking your business processes. In order to make these more practical, let us consider some tangible steps we might take to build our resilient posture—starting with things you could do today.

## 12.6. **Taking Action**

A first step toward making your business more resilient is to come up with a resilience response plan. The idea is to limit losses, keep your business in operation, and accelerate recovery—not just from cyber problems but from a broad range of changes and events that inevitably happen. Besides a disaster, such as fire or earthquake, some of the most urgent problems for your business may be related to a cyber event because an attack could corrupt databases, drain your financial account, or create some other havoc extremely quickly. Planning ahead of time offers the opportunity to think through immediate steps to be taken with the benefit of time and relative calm. This saves critical time, reduces stress, and improves your effectiveness.

In order to create a resilience-oriented plan, keep in mind the health mindset described above. What does it take to keep your business from being vulnerable to every threat that crops up? Start by thinking in operational terms—looking at business functions and their criticality, not just how the technology systems work. What important internal and external dependencies exist, and what key processes are needed to continue operating? You also want to consider a variety of situations that could disrupt your business and ensure your plan covers predictable events. More importantly, though, war-gaming how your business processes would respond to different circumstances can help you to identify what problem indicators would be worth watching and, perhaps, what generic response measures may be effective for different situations.

If you are reading this, your company probably depends upon computer (cyber) technologies in some way. Given your understanding of business processes, how do cyber and other systems factor into business continuity and performance? Are there critical time-sensitive communications? Do you depend upon real-time online financial transactions? Based on this understanding, identify what things need to be kept in operation versus which would be better to shut down in an urgent situation. Shutting off power to the computer or calling the bank's hotline would be akin to first aid principles: check for severe bleeding, breathing, and heartbeat first; you generally have more time to deal with other problems.

Next, consider what information or coordination you would need in order to limit losses and recover important capabilities. Especially in the case of cyber events, it can be difficult to assess actual conditions—of bank accounts, security systems, or other aspects of your business—in order to make decisions and take action. What redundant or alternative resources could be mobilized? Do you have local or cloud-based data backup? Are there others you would need to work with, such as suppliers, service providers, or vendors, and what would you expect them to do? Any of these insights also would be helpful to unpack and coordinate ahead of time, when it is easy to print out phone numbers, checklists, and instructions.

It may be handy to keep paper copies of your plans, checklists, and phone numbers. If you are not a sole proprietor, make sure someone else is familiar, has access to copies, and is ready to step up in your absence. Passwords are a particular challenge because they need to be protected from compromise, but may be essential to restoring services. Make sure you can retrieve them even in the event of computer system problems.

Technology systems, plans, and particularly any paper documents notoriously become outdated. Consider developing a monthly (or maybe quarterly) checklist to trigger a review of things like passwords and patches, accounts, or contacts and keep notes on what you do. Some online systems require or recommend a frequency for password changes; a regular review can help to keep track of which were changed when. Likewise, do a periodic review of your response plan. Have you changed service providers or has someone changed their problem-reporting system? It may be time to update and reprint.

Finally, make it a habit to learn from events and new information. Even relatively small glitches such as network outages can help you to understand vulnerabilities and find work-arounds that might be useful to build into your regular procedures, your technology systems, and/or your resilience plan. Continually watch for sources of information that could provide early warning or indications of important trends. It can be useful to include someone else (internal or external to the company) in the process to open up the thought process.

## 12.7. Conclusions

Cybersecurity programs are important to protect you and your business. Like washing your hands, this is a hygiene issue. Resilience thinking allows you to further reduce the negative impacts of situations that might occur, be they nefarious or random. Moreover, a resilience approach can improve your business performance over time by making sense of what you do, allowing you to improve processes over time. Do not be an intimidated parent—run your business, keep your eyes open, and proactively adjust things to cultivate your competitive advantage.

# 13

# Cybersecurity for Entrepreneurs... and Beyond

Gloria D'Anna and Zachary A. Collier

## 13.1. So What Have We Learned in this Book?

Just like a crystal ball, we are now looking into Peter the Salesman's, turned Peter the Entrepreneur's, future. If Peter the Entrepreneur heeds all of the warnings in this book and implements good cyber hygiene, we will expect him not to have a bad cyber day.

Entrepreneurs need to use good cyber practices and so do small companies and so does everyone across the entire supply chain. If everyone does their part, cybersecurity will get better and better. But everyone must remain adamant and vigilant about cybersecurity. Cybersecurity, as we have learned, is not just for big corporations and governments. Cybersecurity is everyone's responsibility.

Hopefully you have found this book to be both informative and interesting. Our goal was to write this book in a way that allowed you to learn a lot, but with a more "fun" spin. We did not want this to be a boring engineering overview of cybersecurity for entrepreneurs. We really wanted entrepreneurs to pick up this book and implement what is inside.

Mitigating cybersecurity risk is not such an easy thing. An entrepreneur needs to continue to be vigilant. Bad guys may want your company's data or your customers' data or your suppliers' data or even your banking information to utilize for their own financial gain. But you do not have to let the bad guys win. By implementing the lessons you have learned in this book, you should be well on your way to protecting yourself from cyber risks.

Let us take a look back at where we have been in this book.

*Chapter 1:* We started out by being introduced to Peter. Peter was a salesman, who continually had a bad cyber day, and almost died. But Peter was able to cheat death and move onto a new career as an entrepreneur. Peter is a lucky guy. We also learned about how cybersecurity is kind of like sunscreen—it offers an important layer of protection.

*Chapter 2:* Here, Peter is called out by the angel and the devil. The angel wants him to utilize good cyber practices and helps him get resources to start learning. The devil just digs his heel into Peter's shoulder and continually tells him to "do nothing!", "it's too expensive," or, even worse, gives him bad advice. The angel and devil argue with each other. And Peter does not understand exactly what they are saying. He understands bits and pieces. The devil's explanations seemed easier to understand, but the angel's suggestions seem safer. Luckily, this book *Cybersecurity for Entrepreneurs* lands in Peter's hands. And, he starts reading…

*Chapter 3:* In this chapter, Peter learns to secure the communications, including his phone, his email, and his website. Simple things. But they are important for entrepreneurs since communication is pivotal to business.

*Chapter 4:* Peter learns to secure his credit card payments and bank account. He learned about things like encryption and backing up his data. He also learned about how to use cloud services safely. Even a "normal person" would learn excellent methods from just reading Chapters 3 and 4 for their own personal needs.

*Chapter 5:* Peter learns how to utilize a Virtual Private Network (VPN) when he is online. A VPN is simply explained.

*Chapter 6:* As Peter's business grows, he adds Internet-of-Things (IoT) devices to his business—such as a smart doorbell camera. Peter learns in this chapter how to keep his IoT devices secure as best as he can.

*Chapter 7:* As Peter thinks about growing his business, he thinks about developing new products that themselves need to be secure. In this chapter, Peter learns about how to secure new product development. Peter shook his head in Chapter 2 when the angel and devil were arguing about product security. Peter begins to understand the importance of secure product development, even if he does not necessarily understand all of the technical details.

*Chapter 8:* In this chapter, Peter steps back in a more thoughtful mode and thinks about entrepreneurship through the lens of the digital age. He learns about some entrepreneurial strategies and tools to help him better plan and launch his business, with an eye toward cybersecurity during the process. He also learns about the value of networking and utilizing the resources that are out there to help entrepreneurs just like Peter.

*Chapter 9:* The last thing Peter needs is a fine or a lawsuit. In this chapter, Peter learns about the broad contours of cyber law and the various regulations and requirements that he needs to comply with in order to keep customer data secure. He also learned the importance of finding a good lawyer.

*Chapter 10:* As Peter kept reading, he started to look at cyber economics, answering the question "How much should I spend on cyber for my business?"—always a relevant consideration when you are resource-constrained as a startup. Peter learned about various economic metrics that even he could easily calculate to determine which cybersecurity measures deliver the biggest bang for the buck.

*Chapter 11:* Peter learned about cyber insurance in this chapter. Peter does not work in the insurance field so he was a little confused about terms like premiums and deductibles. He initially did not even know that cyber insurance was a real thing. But by the end of the chapter, Peter had learned about why he might need cyber insurance for his business, what it covers, how to get coverage, and some of the important provisions of his policy.

*Chapter 12:* And then of course, Peter learns about what to do when he is hacked. Importantly, he learns about the value of resilience, which is the ability to bounce back quickly after a disruption. Peter learns about the importance of having an incident response plan and backups in case something goes wrong.

## 13.2. Epilogue: Peter Looks toward the Future

Peter received a copy of this amazing book from the angel in Chapter 2 and puts the cyber principles into practice in his business. He becomes conscious that there may be hackers who want to take advantage of his business finances, banking, personal information, etc.

Peter steps back and looks at his business. He surveys his cybersecurity practices and declares that all is good.

But like any good "whack-a-mole game," there can be a cyber incident lurking. Peter needs to be vigilant.

Peter then decides that, yes, it is time to grow his business.

He takes his cybersecurity practices to heart. He goes back to Chapters 7 and 8 and rereads about cybersecurity for product development and entrepreneurship in a digital world. He uses his knowledge about securing his communications, financial transactions, and IoT devices. He does not surf the net on insecure Wi-Fi connections anymore and uses a secure VPN. He integrates cybersecurity into his company's business processes and strategies, including determining how much to spend on cybersecurity solutions and cyber insurance. He complies with relevant cyber laws. His organization becomes resilient.

**FIGURE 13.1**   Peter is very happy. His business continues to grow because he read the book.

Illustrated by Phillip Wandyez.

And, eventually, his company grows. He starts hiring more employees and needs to figure out things like cybersecurity training. How should Peter handle remote workers? What type of security measures should he put in place on company-issued devices? Does Peter need to worry about the cybersecurity practices of his vendors and supply chain partners? What kind of policies and plans does his company need to establish and enforce?

And so Peter sets out to find Volume II of this book.

What is next for Peter? Peter now sells a t-shirt on his website. It is black with white writing. On the front, it says "CYBERSECURITY." On the back, it says "WHAT WOULD PETER DO?"

Perhaps that is something worth thinking about the next time you have a cyber decision to make—CYBERSECURITY: WHAT WOULD PETER DO?

*"Have a Safe Cyber Day!"—Zach and Gloria*

# About the Authors

## Gloria D'Anna

Courtesy of Gloria D'Anna.

Illustrated by Phillip Wandyez.

**Gloria D'Anna** is an engineer, entrepreneur, and multiple patent award holder—an expert in vehicle engineering and cybersecurity. She likes to solve problems, from future light commercial vehicles, to beefing up the cybersecurity of vehicles, always rolling out new tech, reducing inefficiencies, and driving business. Her latest obsession is CyberPhysical Security—or as she described is, "Where Are You in the Universe?"

She began her career at GM, winning an MBA Fellowship to the University of Chicago, moving on to Ford, Navistar, Textron, Eaton, and Ricardo. Later, she led sales at three successful startups, addressing challenges from school safety to building connected devices for law enforcement. She currently works at Ford on future vehicles, and is known as "The Antenna Lady."

Gloria has been working with SAE for the last decade, creating and moderating popular and educational cybersecurity technical sessions from Commercial Vehicle markets to the Internet of Things, and now SAE Aerotech. Her 2018 book, *SAE Cybersecurity for Commercial Vehicles* focused on Medium-Duty and Heavy Duty trucks.

She is the recipient of SAE International's 2018 Forest R. McFarland Award for Automobile Electronics Activity.

She is currently the CEO of Greater Telecommunication Systems, LLC, a private woman-owned company—where she is merging several pieces of software (including Artificial Intelligence and Semantic Ontology) and hardware to improve CyberPhysical Systems. She co-chairs SAE's G-32 CyberPhysical Systems with Boeing.

*SAE Cybersecurity for Entrepreneurs* is her second book with SAE International.

# Zachary A. Collier

Courtesy of Zachary Collier.

Illustrated by Phillip Wandyez.

**Zachary A. Collier** is Assistant Professor in the Department of Management at Radford University. His research interests include risk analysis and decision analysis, which he applies to problems at the intersection of technological, organizational, and societal domains.

He currently serves as Co-chair of the National Defense Industrial Association (NDIA) Electronics Division's Trust and Assurance Committee. Dr. Collier is a Fellow of the Center for Risk Management of Engineering Systems at the University of Virginia and a Visiting Scholar at the Center for Hardware and Embedded Systems Security and Trust (CHEST). He has held various leadership positions within the Society for Risk Analysis, such as President of the Decision Analysis and Risk Specialty Group and President of the Resilience Analysis Specialty Group. He serves on the INFORMS (Institute for Operations Research and the Management Sciences) Advocacy Governance Committee.

Dr. Collier has contributed as a risk management subject matter expert to the development of industry standards through SAE International, both in the G32 and G19A committees.

He is Managing Editor of the Springer journal *Environment Systems and Decisions* and is a member of the Editorial Board of *Risk Analysis*.

He earned his PhD in Systems Engineering from the University of Virginia, a Master of Engineering Management from Duke University, and a Bachelor of Science in Mechanical Engineering from Florida State University.

# Simon Hartley

Courtesy of Simon Hartley.

**Simon Hartley** is a veteran of successful startups, small businesses, and Fortune 100 companies and a subject matter expert in the business of cybersecurity. He leads United States (US) Sales and Business Development for US government and critical infrastructure at the startup Quantinuum, leveraging quantum technology to defend against cybersecurity threats and deliver new ways to solve scientific problems.

Previously, he introduced a new mobile security platform with startup CIS Mobile, after earlier working with Apple and Samsung to harden their platforms for US government security. Prior to that, Simon was Vice President (VP) of Business Development / Co-founder of Internet of Things cybersecurity startup RunSafe Security, VP of Sales and Marketing at mobile cybersecurity startup Kaprica Security (now Samsung), and Director of Worldwide Sales and Marketing, restarting small business Thursby Software (now Identiv) in mobile security.

Simon began his career in nuclear software engineering in England and France. He then held executive roles at HP, Red Hat, and Capgemini. Past end customers cover government, finance, healthcare, transport, and energy. He is a Certified Information Systems Security Professional (CISSP) and holds a Bachelor of Science in Physics, Master of Science in Law and Cybersecurity, and Master of Business Administration degrees.

# Chris Sundberg

Courtesy of Chris Sundberg.

**Christopher Sundberg**, GICSP is a Product Cybersecurity Engineer in the Corporate Technology Office for Woodward, Inc. Mr. Sundberg is responsible for security architecture, secure development lifecycle, and product security compliance across a wide variety of cyberphysical devices sold by Woodward, Inc.

Mr. Sundberg's work history spans over 30 years concentrating on wireless communication and embedded systems development.

His background includes cyberphysical systems, industrial control networks, vehicle networking, and emerging digital technologies such as artificial intelligence, cloud architectures, and Industrial Internet of Things.

Mr. Sundberg has actively participated in the development of the SAE G-32 Cyber Physical Systems Security standard, serving on the Software Assurance subgroup, curator for the G-32 Weekly Security Items newsletter, and liaison to a number of other SAE committees (SAE G-34/Eurocae WG-114 Artificial Intelligence in Aviation, G-31 Electronic Transactions for Aerospace, and S-18 Aircraft and System Development and Safety Assessment Committee).

# Dennis Vadura

Courtesy of Dennis Vadura.

**Dennis Vadura** is CEO of BADU Networks Inc. He has 40+ years in all aspects of Computing including software development, networking, system architecture and analysis. He is a highly experienced hands-on systems and solution architect, senior software developer, and expert in Systems, Object-Oriented software, programming languages, distributed computing, internet technologies, and network protocols. He has used all of the above skills in the implementation and ongoing support of both large-scale and small-scale software systems including servers, proxies, and GUI applications including cloud, mobile, desktop, and web services.

At BADU Mr. Vadura is responsible for a team that developed a highly scalable network TCP proxy using optimized TCP data flows, 20 million TCP sessions (40 million connections) and 100k+/second active data flows on a single server. He has also developed both Android and iOS custom applications, as well as a VPN that uses the WarpEngine TCP proxy as its backbone core infrastructure including its Android App.

Mr. Vadura personally wrote proxy code in C++ with kernel drivers (in C) to support SDWAN and VPN services. He also fixed over 20 network related bugs in the linux TCP/IP stack. Created a custom framework to simplify C++ concurrency and networking primitives including sockets, sockaddrs, and various related threading libraries.

Mr. Vadura deployed a VPN to cloud services (nodes all over world) with focus on WarpEngine for scaling VPN data flows, REDIS keystore cache for distribution of credential data, and haproxy for a scalable front-end login subsystem (millions of hits of capacity per day). As part of the deployment he implemented a private Certificate Authority supporting certificate issuance and revocation.

As part of deploying VPNs, and during 2022, Mr. Vadura spent a significant amount of time analyzing various consumer VPN options. A key part of the analysis was a realization that the VPN review sites provide a set of easily accessible comparison metrics with most being of no value in assessing the privacy preserving potential of a VPN.

# Peter Laitin

Courtesy of Peter Laitin.

**Peter Laitin** is a specialist in enterprise software sales, focused on the intersection of cybersecurity, mobility, Internet of Things, and cloud for the Department of Defense, civilian government, healthcare, and financial verticals with seasoned and startup Information Technology companies.

He has worked with several of the nation's top cybersecurity firms, including RSA Security (acquired by EMC), Verisign (acquired by Symantec), Invincea Inc. (acquired by Sophos), Thursby Software Systems (acquired by Identiv), and others.

Peter helped grow revenue for these companies by developing strategic partnerships in their target markets. His expertise in emerging technologies enabled him to identify new markets that needed more maturity to develop their own sales teams but were large enough to generate millions of dollars in sales.

He also assisted these companies with product positioning, helping them understand how best to position themselves against competitors so they could win business from customers looking for solutions to specific problems.

Peter joined a growing family of DC-based cybersecurity and mobility startups, Kaprica, RunSafe, and Tachyon, all focused on helping organizations better find their most valuable assets – people – and protect them from cyberattacks.

# Kenneth G. Crowther

Courtesy of Kenneth G. Crowther.

**Dr. Kenneth G. Crowther** is the Product Security Leader for Xylem. He was formerly Product Security Leader for General Electric Global Research and Principal Engineer at the MITRE Corporation. He teaches applied quantitative risk analysis at the University of Virginia and Georgetown University, has published dozens of peer-reviewed manuscripts on topics related to risk analysis and homeland security, and served as the Chair of Attack and Disaster Preparedness Track of the Institute of Electrical and Electronics Engineers Homeland Security Technology Conference, as the Assistant Area Editor for the journal *Risk Analysis*, as Chair of the Engineering and Infrastructure Specialty Group and Security and Defense Specialty Group of the Society for Risk Analysis, and on the Board of Directors of the Security Analysis and Risk Management Association. His research and publications in risk analysis have received various honors from the Institute for Information Infrastructure Protection, the International Council on Disaster Research, the University of Virginia Department of Systems and Information Engineering, the Department of Homeland Security, and the Center for Risk Management of Engineering Systems, among others. In addition to his current work at Xylem, he serves on the ISA Global Cybersecurity Alliance subcommittee for IIOT cybersecurity certifications and on a committee of the Military Operations Research Society to train and certify risk analysts for doing national security risk analyses. Dr. Crowther holds a PhD in Systems and Information Engineering from the University of Virginia and a Bachelor of Science in Chemical Engineering from Brigham Young University.

# Dale W. Richards

Courtesy of Dale W. Richards.

**Dale W. Richards** is Founder and CEO of AppCreative, an offshore software development company based in Lehi, Utah, and Tbilisi, Georgia. AppCreative provides seamlessly integrated offshore development teams to help software-as-a-service companies accelerate their product roadmaps. Dale is passionate about building applications (apps) and inspiring others to do the same. He vlogs about app creation on YouTube, where he shares app ideas, how-tos, and industry news. He is an international keynote speaker and was the closing keynote for the International Project Management Conference in Moscow, Russia, in December 2019. Dale received his Master in Business Administration degree from the James Madison University College of Business and has held a Project Management Professional certification since 2009.

# Samantha Bryant Steidle

**Dr. Samantha Bryant Steidle** is an Entrepreneurship Instructor and the Director of the Venture Lab at Radford University. She is also a contractor for the US Department of Treasury's 2021 State Small Business Credit Initiative (SSBCI), which provides a combined $10 billion to states, the DC, territories, and tribal governments. SSBCI funding provides small businesses access to venture capital and other funding needed to invest in post-pandemic job creation. Dr. Steidle's work history spans over 20 years and includes co-launching six entrepreneurial spaces prior to the Venture Lab: a startup incubator, technology and life-science public-private accelerator, inventor's fabrication laboratory, and three coworking spaces. Through local and national startup thought leadership, Steidle cultivated robust relationships with government officials, funders, technology councils, academic institutions, investors, and other successful entrepreneurs.

# Jennifer Dukarski

Courtesy of Jennifer Dukarski.

**Jennifer Dukarski** is a Shareholder based in Butzel Long's Ann Arbor office, practicing in the areas of intellectual property, media, and technology and focusing her practice on the intersection of technology and communications with an emphasis on the legal issues arising from emerging and disruptive innovation: digital media and content, vehicle safety, connected, electrified and autonomous cars, and data privacy and security. Jennifer was named one of the 30 Women Defining the Future of Technology in January 2020 by Warner Communications for her innovative thoughts and contributions to the technology industry. She is a Certified Information Privacy Professional concentrating on the United States private sector privacy and data protection law (CIPP/US).

# C. Ariel Pinto

**C. Ariel Pinto** an entrepreneur, research director, professor, mentor, and an established author with 100+ published manuscripts, abstracts, and presentations in engineering and risk management, including two textbooks. He has engineering degrees from the University of Virginia and the University of the Philippines and has 20+ years of experience in Systemic Risk Analysis and Management. He recently received the prestigious Fulbright Grant in Cybersecurity to the United Kingdom. He was previously at Old Dominion University and is currently with the Department of Cybersecurity at the University at Albany.

# Luna Magpili

Courtesy of Luna Magpili.

**Dr. Luna Magpili** is an Associate Professor of Engineering and Technology Management at Washington State University and has been involved in academic teaching and research for more than 10 years. She has extensive international experience as an industrial engineer and consultant for various manufacturing and export enterprises and, recently, for insurance and healthcare organizations on risk management and artificial intelligence. She also served as a program officer for International Relief and Development and managed projects in relief work and infrastructure development across the globe. She currently serves as panel reviewer for various programs at the National Science Foundation, National Aeronautics and Space Administration, and Department of Defense. She is a co-author of the book *Operational Risk Management* and several seminal papers on risk technology and complex system governance. She has also served as a referee to various journals such as *Risk Analysis, Engineering Management Review,* and *PLOS One* and currently a member of the International Council on Systems Engineering, Asia–Europe Meeting, the Theory of Constraints International Certification Organization, Risk and Insurance Management Society Inc., and American Society for Engineering Education.

# Howard Miller

**Howard Miller,** CRM (Certified Risk Manager), CIC (Certified Insurance Counselor), CyRP (Cyber Risk Professional), Vice President, Account Executive at Alliant Insurance Services, Inc., Irvine, California, US. He has over 20 years of experience advising clients on custom commercial insurance programs, with focus on cyber insurance, technology risk, and protecting long-term success for clients through risk management and insurance. He was Co-lead of the Risk Management Framework subcommittee for the published standard JA7496 Cyber-Physical Systems Security Engineering Plan, SAE G-32, and author of the enterprise risk management framework published in *Environment Systems and Decisions* "Enterprise risk management optimization (ERMO)."

# Paul E. Roege

Courtesy of Paul E. Roege.

**Paul E. Roege** is a researcher, author, and practitioner in the fields of energy and resilience, with more than 15 papers, articles, and book chapters. He has nearly 40 years of experience as an engineer and leader in engineering, construction, and research, primarily in the energy field. As a United States (US) Army engineer officer, Colonel Roege built military infrastructure and led combat engineering capabilities in Europe, Asia, Africa, and Central America. He planned and coordinated the reconstruction of Iraqi oil production systems in 2003; later, he developed energy requirements and strategies for military operations and was an early advocate within the Department of Defense for resilience as a guiding principle for community and national security. In his civilian career, Colonel Roege led engineering efforts associated with operational management and decommissioning of US nuclear weapons production facilities and disposition of plutonium from US and former Soviet weapons programs. Paul is a registered professional engineer and a West Point alumnus, with graduate degrees from Boston University (Business) and the Massachusetts Institute of Technology (Engineering).

# About the Illustrator

## Phillip Wandyez

**Phillip Wandyez** is a Civil Engineer, Attorney, and is currently pursuing a degree in Computer Science with a focus on machine learning and cybersecurity. In his spare time he enjoys making illustrations.

# Index

## A

Access control policies, 62
Action trumps analysis, 96
ADint, 20
Ad serving, 50
Advanced Encryption
  Standard (AES), 64
Affordable loss, 96
Aggregate limit, 161
Agile, 74, 78, 79
Amazon Web Services
  (AWS), 86
"Anti-phishing" tool, 16
Apache, 82
Apple smartphones, 18, 23
Application programming
  interface (API),
  38–39, 64
Apricorn, 19
Attack surface, 60–61
Automated Clearing
  House (ACH)
  transaction, 39
Automobiles, 131

## B

Backup strategy, 35–36
Banking, 16, 124, 151,
  181, 184, 186. See
  also Financial
  transaction
Berkeley Software
  Distribution (BSD),
  82
Botnet, 62
Brainstorming, 79
Breach costs, 163
Brilliant, Ashleigh, 4–5

Building Security In
  Maturity Model
  (BSIMM), 77, 78
Business continuity, 35–36
Business e-mail
  compromise (BEC),
  32
Business interruption, 164
Business Model Canvas,
  99–100

## C

California Consumer
  Privacy Act (CCPA),
  167
Card Verification Value
  (CVV) code, 39–40
Catastrophic cybersecurity
  incidents
  analysis, 140–141
  first-generation AI
    authentication app,
    141–142
  formula, 140
  pushing, 140
  second-generation AI
    authentication app,
    142–144
  time consideration of
    investments,
    139–140
Center for Internet Security
  (CIS), 19
Certificate Authority
  attacks, 48–49
Children's Online Privacy
  Protection Act
  (COPPA Rule), 108

Chilling effect, 20
Ciphertext, 63
CIS Mobile, 24
Cloud data storage
  API security, 38–39
  credit card processing,
    39–41
  online tools, 36–37
  risks, 37
  SOC reports, 37–38
Cloud security, 21–22
Clover, 40
Code-based analysis, 83
Collier, Zachary A., 3–5
Communication,
  6, 184
  cybersecurity awareness,
    16–17
  e-mail
    limitations, 19–20
    phishing, 21
    security features,
      20–21
  phone, 24
    cars, events, and
      overseas travel, 25
    PACE, 25–26
    recommended
      phones, 23
    voice, text, and
      messaging, 24–25
  PQC, 26
  stakes, 18–19
  web safety
    bad sites and links, 23
    cloud security, 21–22
    web security, 22
    web tracking, 22–23

Confidentiality, integrity,
and availability
(CIA), 15–16, 152
Confidentiality protection,
131
Consumer-off-the-shelf
(COTS) devices,
23, 24
Contingent/dependent
system, 164
Continuous integration,
continuous
development (CICD)
pipelines, 80, 84
Cost center
*vs.* profit center, 130–132
RROI (*see* Risk-based
return on
investment (RROI))
Coverage, 155, 162
COVID-19, 51, 97
Crazy quilt, 96
Credit card processing,
39–41
Crises management, 163
Critical infrastructure,
113, 114
Crowd-source, 84
Cryptographic hash, 66
Customer development,
100–101
Cyberattack, cost of, 4
Cyberattack
ramifications, 98
Cybercriminal, 2, 158
Cyber economics, 185
cost center *vs.* profit
center, 130–132
costs, benefits, and
information,
144–145
definition, 127
delayed NPV and
catastrophic
cybersecurity
incidents
analysis, 140–141
first-generation AI
authentication app,
141–142

formula, 140
pushing, 140
second-generation AI
authentication app,
142–144
time consideration of
investments,
139–140
laptop computer, 127–128
ROI, 132–134
RROI
baseline scenario, 135,
137–138
cost, 135
formula, 135
implementation, 134
incident risk, 137–138
incident type, 136
net benefits, 138–139
net bypass rate,
136–137
residual risk, 135
technologies and
policy-related
investments, 135
value of your product/
service, 128–130
Cyber extortion, 164–165
Cyber-infrastructure, 131
Cyber insurance, 6,
185–186
agents, 153
applying for, 153–155
backup and recovery,
157–158
bonus, 149–150
definition, 151–152
email security, 155–156
exposures, 148, 149
first-party insuring
agreements
breach costs and
crises management,
163
business interruption,
164
cyber extortion,
164–165
cyber risk, 162–163
growth mode, 152–153

insurance policy
contractual aspects,
160
paying attention,
160–162
internal security controls,
156–157
losses, 149, 150
network security
vulnerabilities, 155
online educational
portals and technical
resources, 150
phishing, 158–159
plans and programs, 150
renewal, 153
residual risk, 149
risk management, 147, 148
third-party insuring
agreements
legal exposure, 165
media liability, 166
network security and
privacy, 166
regulatory defense and
penalties, 167
technology errors and
omissions, 166–167
Cyber law, 6, 185
business impact, 123–124
data privacy, 107
European Union and
International
Requirements,
121–123
federal laws and
regulations
FTC Act, 108–110
GLBA, 110–111
HIPAA and HITECH
Act, 111–115
state laws and
regulations
data breach laws,
115–118
minimum standard,
118–121
reasonable data
security measures,
118–121

Cyber liability policy,
    160, 161
Cyber professionals, 94
Cyber resilience, 6, 186
    absorption and
        responding, 176
    advantage, 177
    anticipation, 176
    challenge of parenting,
        171–172
    definition, 170–171
    design principles
        adaptability, 179
        dependencies, 179
        diversity, 179
        flexibility, 179
        margins/buffers, 178
        resistance, 178
        sensing and
            information, 178
        stability, 178–179
    first aid principles,
        180–181
    holistic approach, 1
        73–175
    information/
        coordination, 181
    learning and adaptation,
        177
    network outages,
        181–182
    "no-brainer" approach,
        172
    planning, 180
    problem-reporting
        system, 181
    protection vs.
        performance,
        169–170
    recovery, 176–177
    "whack-a-mole" situation,
        172
Cybersecurity and
    Infrastructure
    Agency (CISA), 19
Cybersecurity terms, 124

**D**

D'Anna, Gloria, 3–5
Dashlane, 18

Data breach, 4
    communication, 16
    financial transaction,
        32–33
    laws
        notification, 116–117
        penalties, fines, and
            possibility, 118
        personal information,
            116
        reporting, 117–118
        security incident/
            breach, 116
The Data Breach
    Notification Rules,
    112–113
Data flow diagram (DFD),
    78, 79
Data privacy, 64, 107, 115
Data Protection Regulation
    (GDPR), 167
Debug interfaces, 61
Decryption, 63
Defense-in-depth, 156
Delayed Net Present Value
    (NPV)
    analysis, 140–141
    first-generation AI
        authentication app,
        141–142
    formula, 140
    pushing, 140
    second-generation AI
        authentication app,
        142–144
    time consideration of
        investments,
        139–140
Delivery of service, 129
Denial-of-service attacks,
    62
Department of Homeland
    Security, 113
Deployment mechanism, 84
Design principles, cyber
    resilience
    adaptability, 179
    dependencies, 179
    diversity, 179
    flexibility, 179

margins/buffers, 178
    resistance, 178
    sensing and information,
        178
    stability, 178–179
Design thinking, 96–97
DevOps, 74, 78, 79
Diffie-Hellman key
    exchange, 48
Digital products/services.
    *See* Product security
Digital signatures, 64
Digital surveillance, 20
Dynamic application
    security testing
    (DAST), 83

**E**

Economics, 6
Effectual reasoning/
    discovery-driven
    planning, 98
Effectuation, 96
E-mail security, 155–156
    limitations, 19–20
    phishing, 21
    security features, 20–21
Empathy-driven process, 96
Encryption, 18, 26, 35, 47,
    52, 63, 64, 68, 111,
    112
End-to-end encryption
    (E2E), 20, 24
Entrepreneurial ecosystems,
    101–103
Entrepreneurial
    networking,
    101–103
Entrepreneurial planning, 6
Entrepreneurial thinking,
    98–99
Entrepreneurship
    education, 3–4
Ethical hacking, 84
European Union and
    International
    Requirements,
    121–123
Executive Orders, 113–115
Exposure, definition, 148

## F

Fair Credit Reporting Act (FCRA), 108–109
Federal laws and regulations
FTC Act, 108–110
GLBA, 110–111
HIPAA and HITECH Act, 111–115
Federal Trade Commission Act (FTC Act), 108–110
FileVault2, 35
Financial services and applications, 37
Financial transaction, 6, 184–185
business continuity, 35–36
cloud and SaaS
API security, 38–39
credit card processing, 39–41
online tools, 36–37
risks, 37
SOC reports, 37–38
data breach, 32–33
data encryption, 35
IAM, 34–35
policies and procedures, 34
security controls, 34
Fingerprinting, 22–23
Firefox browser, 22
First-generation artificial intelligence (AI) authentication app, 141–142
First-party insuring agreements
breach costs and crises management, 163
business interruption, 164
cyber extortion, 164–165
cyber risk, 162–163
Flight malfunction, 5
*FTC v. Sandra Rennert*, 109
Fuzz testing/fuzzing, 83–84

## G

General cognitive ability, 98
General Data Protection Regulation (GDPR), 121, 122
Glacier Security, 18–19
GoDaddy, 22
Google Gmail, 20–21
Gramm-Leach-Bliley Act (GLBA), 109–111

## H

Hackers, 16, 23, 37, 64, 67, 68, 130, 140
Health Information Technology for Economic and Clinical Health Act (HITECH Act), 111
the Data Breach Notification Rules, 112–113
Executive Orders, 113–115
the Privacy Rule, 111–112
the Security Rule, 112
the Transactions Rule, 113
Health Insurance Portability and Accountability Act (HIPAA), 167
the Data Breach Notification Rules, 112–113
Executive Orders, 113–115
the Privacy Rule, 111–112
the Security Rule, 112
the Transactions Rule, 113
Healthy living, 173–174
HTTPS Everywhere, 22
Human-centered solution/ intervention, 96–97
Hypertext Transfer Protocol Secure (HTTPS), 51
ad serving, 50
business use, 47
IoT, 63

location and information privacy, 50
man-in-the-middle attack, 48–49
SSL, 47–48
TLS algorithm, 47–48
tracking cookies, 50
traffic-based attacks, 49–50
webpage information, 47

## I

IBM/Ponemon report, 163
Iceberg Model of Systems Thinking, 97–98
Identity and access management (IAM), 34–35
Incident command system, 88
Incident response (IR), 32–33
Industry 4.0, 131
Information Technology (IT), 130
*In re BJ's Wholesale Club, Inc.*, 109–110
Inscape, 56–57
Insider threats, 37, 38
Insurance policy, 160
contractual aspects, 160
first-party insuring agreements, 162–165
paying attention, 160–162
Intellectual property (IP), 16, 166
Internal Controls over Financial Reporting (ICFR), 37
Internal security controls, 156–157
International Society of Automation (ISA) Secure Development Lifecycle Assurance (SDLA), 77
Internet communication, 32–33

Internet of Things (IoT),
    121, 185
anti-virus software, 68
attack surface, 60–61
audit actions, 68
authentication tools, 68
built-in firewalls, 66
cryptographic hash, 66
DDoS attack, 66
device infection, 68
encryption, 68
logging/alerting
    functionality, 67
Man-in-the-Middle
    attack, 63–65
privacy protection, 67
software updates and
    patches, 61–62, 68
vendor's security, 67
Internet of Things Security
    Recommendations
    (INT 1–12), 68
Internet protocol (IP)
    address, 50
Internet service provider
    (ISP), 26, 49, 50
ISA Global Cybersecurity
    Alliance, 88

**K**
Key performance indicators
    (KPIs), 38
"Know Your Client" (KYC)
    types, 17

**L**
Lean Startup, 100
Least privilege, 156
Lei Geral de Proteçao de
    Dados Pessoais
    (LGPD), 122
Lemonade principle, 96
Log4j, 88
Lookout, 19, 23

**M**
Malware, 23, 25, 62, 137,
    139, 155
Managed service providers
    (MSP), 36
Man-in-the-Middle attacks,
    48–49, 63–65

Manufacturing of products,
    129
Massachusetts Institute of
    Technology (MIT), 82
Massachusetts state security
    law and regulations,
    120–121
Media liability, 166
Microsoft Outlook, 20–21
Microsoft's BitLocker, 35
Minimum standard and
    reasonable data
    security measure
    laws, 118
California Internet of
    Things Law, 121
data security obligations,
    118–119
Massachusetts state
    security law and
    regulations, 120–121
NYDFS, 119–120
Minimum viable product
    (MVP), 71–72
MITRE ATT&CK model,
    88–89
MobileIron, 24
Multi-Factor
    Authentication
    (MFA), 34–35

**N**
National Institute of
    Standards and
    Technology (NIST),
    114
National Institute of
    Standards and
    Technology (NIST)
    Secure Software
    Development
    Framework (SSDF),
    76, 78
Netflow data, 49–50
Network security, 64, 88,
    136, 155, 166
New York Department of
    Financial Services
    (NYDFS), 119–120
NordVPN, 52

**O**
Office of Foreign Assets
    Control (OFAC), 164
Open-source tools, 83
Open Systems
    Interconnection
    (OSI) network layer,
    54
Open Worldwide
    Application Security
    Project (OWASP)
    Software Assurance
    Maturity Model
    (SAMM), 76–78
Operational Technologies
    (OT), 131
OSint, 20
OWASP Application
    Security Verification
    Standard, 39

**P**
1Password, 18
Password security, 10–12
Patching software
    vulnerabilities, 156
Payment Card Industry
    Data Security
    Standard (PCI DSS),
    40, 167
PayPal, 40
Penetration testing, 84
Permissive software license,
    82–83
Personally identifiable
    information (PII),
    34, 116
Phishing, 21, 32, 158–159
Phone security, 24
cars, events, and overseas
    travel, 25
PACE, 25–26
recommended phones, 23
voice, text, and
    messaging, 24–25
Pixel smartphones, 18, 23
Plaintext, 63
Point-to-Point Tunneling
    Protocol (PPTP),
    44–45

Post Quantum
Cryptography
(PQC), 26
Pretty Good Privacy (PGP),
20–21
Primary, Alternate,
Contingency, and
Emergency (PACE)
communication,
25–26
Privacy Badger, 22
Privacy liability, 166
The Privacy Rule, 111–112
Product security, 6, 12–13,
185
activities, 76
security flaw
Agile and DevOps, 74
costs of fixing flaws, 73
data breach victims, 73
product/integration
team, 72
standards, 76–77
sustainment, 75
documentation and
training, 85–86
PSIRT, 86–88
third-party component
monitoring, 88–89
vulnerability
management, 88–89
warranty requests, 85
testing, 75, 85
DAST, 83
delayed product
release, 82
fuzz testing/fuzzing,
83–84
penetration, 84
regression, 84
SAST, 83
SCA, 82–83
tools, 81–82
threat modeling, 75
Agile and DevOps
frameworks, 78, 79
design, architecture,
and
communications, 78
features, 77–78

NIST SSDF model, 78
potential weaknesses,
77
process of, 79–81
Product security incident
response team
(PSIRT), 86–88
Profit center, 130–132
Protected health
information (PHI),
111–113
Proton Mail, 21

**R**
Ransomware, 16, 37, 164
Regression testing, 84
Repeatable and scalable
business
model, 100
Research and development
(R&D), 130
Residual risk, 135
Return on Investment
(ROI), 132–134
Return on Security
Investment (ROSI),
135
Risk-based return on
investment (RROI)
baseline scenario, 135,
137–138
cost, 135
formula, 135
implementation, 134
incident risk, 137–138
incident type, 136
net benefits, 138–139
net bypass rate,
136–137
residual risk, 135
technologies and
policy-related
investments, 135
Risk-informed decisions, 3
Risk management, 147, 148
Risk mitigation, 184

**S**
Samba, 56–57
Samsung smartphones,
18, 23

Second-generation artificial
intelligence (AI)
authentication app,
142–144
Secure Shell (SSH), 61
Secure Sockets Layer (SSL),
22, 47–48, 63, 109
Security controls, 75, 85
DAST, 83
delayed product release,
82
fuzz testing/fuzzing,
83–84
penetration, 84
regression, 84
SAST, 83
SCA, 82–83
tools, 81–82
Security incident/breach,
87, 116
Signal, 18, 24
Sites and links, 23
Slack, 25
Small Business
Development
Centers and
Technology
Councils, 103
Social media
authentication tools, 68
marketing for company,
10–11
security questions, 21
Software-as-a-Service
(SaaS)
API security, 38–39
credit card processing,
39–41
online tools, 36–37
risks, 37
SOC reports, 37–38
"Software bill of materials"
(SBOM), 82
Software composition
analysis (SCA),
82–83
Software performance, 157
Software security tools,
16–17
SolarWinds, 88

South Korea's Personal
   Information
   Protection Act, 123
SpaceX, 26
SPAM function, 2
Spoof, Tamper, Repudiation,
   Information
   Disclosure, Deny
   Service, or Elevating
   Privilege (STRIDE),
   79, 80
Square, 40
The Standards for
   Electronic
   Transactions and
   Code Sets, 113
StartCom, 48–49
Startup process, 185
   cyberattack, 94
   cyber risks, 94
   modern entrepreneurial
      networking
      ecosystems, 104–103
   resources, knowledge,
      and skills, 103–104
   modern entrepreneurial
      strategies
   design thinking, 96–97
   effectuation, 96
   entrepreneurial
      thinking, 98–99
   resources, knowledge,
      and skills, 103, 104
   systems thinking,
      97–98
   modern entrepreneurial
      tools
   Business Model
      Canvas, 99–100
   customer development,
      100–101
   Lean Startup, 100
   resources, knowledge,
      and skills, 103–104
State laws and regulations
   data breach laws, 115–118
   minimum standard,
      118–121
   reasonable data security
      measures, 118–121

Static application security
   testing (SAST), 83
Stolen credentials, 33
Subscriber identity module
   (SIM)-swapping
   attacks, 18
System and organization
   controls (SOC),
   37–38
Systems thinking, 97–98

T
Target architecture model,
   68
Teams, 25
Telnet, 61
Third party access, 62
Third-party component
   monitoring, 88–89
Third-party insuring
   agreements
   legal exposure, 165
   media liability, 166
   network security and
      privacy, 166
   regulatory defense and
      penalties, 167
   technology errors and
      omissions, 166–167
Threat modeling, 75
   Agile and DevOps
      frameworks, 78, 79
   design, architecture, and
      communications, 78
   features, 77–78
   IoT devices, 62
   NIST SSDF model, 78
   potential weaknesses, 77
   process of, 79–81
   to VPN, 52
TikTok, 23, 56
Tor, 22–23
TorGuard, 52
Traffic-based attacks,
   49–50
Traffic hijacking, 62
The Transactions Rule, 113
Transport Layer Security
   (TLS), 35, 47–49
Trusted Protection Module,
   35

Trust Services Criteria,
   37–38
Two-factor authentication
   (2FA), 11–12,
   18, 68

U
User authentication, 62
User behavior analytics, 156

V
Verizon Data Breach
   Investigation
   Report, 163
VikingVPN, 52
Virtual Private Network
   (VPN), 185, 186
   communication,
      18–20
   customer data, 46
   features, 53–54
   HTTPS, 51
      ad serving, 50
      business use, 47
      location and
         information privacy,
         50
      Man-in-the-Middle
         attack, 48–49
      SSL, 47–48
      TLS algorithm,
         47–48
      tracking cookies, 50
      traffic-based attacks,
         49–50
      webpage information,
         47
   Internet surfing, 51–52
   overview, 43–44
   PPTP, 44–45
   privacy-focused VPN,
      54–55
   properties, 44–45
   TikTok, 56
Voice Over Internet
   Protocol (VOIP)
   calls, 24
Vulnerability
   Agile and DevOps, 74
   attack surface, 60–61
   costs of fixing flaws, 73

data breach victims, 73
gain access, 52
management, 62, 88–89
network security, 155
patching software, 156
product/integration
    team, 72
PSIRT, 86–88

**W**
Web safety
    bad sites and links, 23
    cloud security, 21–22
    web security, 22
    web tracking, 22–23
Wickr, 18, 24
Workspace One, 24

Written information
    security policy
    (WISP), 110

**Z**
Zero-trust architecture, 88
Zimperium, 19, 23
Zoom, 25

Printed in the United States
by Baker & Taylor Publisher Services